# The 5Es of Inquiry-Based Science

The
5Es of Inquiry-Based
Science

**Authors**
Lakenna Chitman-Booker, M.A.Ed.
Kathy Kopp, M.A.

SHELL EDUCATION

## Publishing Credits

Dona Herweck Rice, *Editor-in-Chief*; Robin Erickson, *Production Director*;
Lee Aucoin, *Creative Director;* Sara Johnson, M.S. Ed., *Senior Editor;*
Leah Quillian, *Assistant Editor;* Grace Alba, *Cover Designer;*
Corinne Burton, *M.A.Ed., Publisher*

## Shell Education

5301 Oceanus Drive
Huntington Beach, CA 92649-1030
http://www.shelleducation.com

### ISBN 978-1-4258-0689-7

© 2013 Shell Educational Publishing, Inc.

# The 5Es of Inquiry-Based Science

# Foreword

For many students, science took a backseat when No Child Left Behind was implemented. However, the subject has been brought back into focus through the science literacy standards included in the Common Core State Standards and the Next Generation Science Standards. School administrators are taking note that their teachers and students need renewed support and encouragement if they are to be competitive in a global society. Now, there has never been a more important time in education for teachers to engage students in meaningful science instruction. Likewise, there is not a more respected and proven method for doing this well than the 5E instructional model for science inquiry.

During my career in science education, I was always willing to try new ideas, lessons, and strategies—with varying degrees of effectiveness. There was a lot of trial and error in my early classroom years; however, once I became aware of the 5E model, I began to see more and more student success.

*The 5Es of Inquiry-Based Science* will help professional development providers, administrators, and teachers understand how the 5E model works. Furthermore, the tools needed to make the 5Es a regular part of instruction are readily available. Teachers who use the 5E model of instruction will begin to see students take more responsibility for their own learning. They will be more engaged in learning activities than ever before. Learning will be scaffolded as students are highly engaged in interesting activities that result in problem solving and, in many cases, teamwork.

As you will discover, the constructivist view of education forms the foundation for the 5E model. In this view, students build on what they already know and understand. Misconceptions are reformed into proper conceptions as students become explorers involved in hands-on science. It has been my experience that once teachers see this in action, they are as hooked on science as their students are. In my mind, I envision a teacher trying some of the strategies in this book and experiencing success—the kind of success that makes him or her want to pull out this resource again. Whether it is to find another way to "engage" students in a particular topic or to use one of the many methods to help students "explore" science—this book is the type of resource that will be bookmarked, filled with sticky notes, highlighted, and eventually "dog-eared." My favorite teacher resources look like this, and they are the ones I refer to over and over again and never tire of.

No matter where you are in your career as an educator, I suspect you will find that this book makes you think in ways that you have not considered previously about how to share science with students. For me, it certainly laid out the 5E model in easy-to-understand terms so that I would feel comfortable sharing with new teachers and education veterans. In the world of science education, we are constantly working with a changing target—whether that target is new standards, new discoveries in science, or administrators' expectations. This book, like the 5E model itself, will stand the test of time and continue to help teachers boost students' understanding and achievement in science.

—Steve Rich
Science Center Director
University of West Georgia
NSTA Director of Professional Development

# An Overview of the 5E Instructional Model

*"Science is a way of thinking much more than it is a body of knowledge."*

—Carl Sagan (1986)

Today's science teachers face many challenges. Some challenges, such as poor student attitudes, are time-honored favorites. Other challenges, such as shifts in "the best" content and delivery methods, are recycled dilemmas, renewed and refreshed for the 21st century. Yet other challenges, such as high-stakes tests and society's need for a highly specialized workforce, challenge today's teachers like no other time in the history of education. As times continue to change, classroom teachers must focus their instructional delivery methods to meet the demands of challenging curriculum (content standards) and learning expectations (critical thinking and problem solving); compete against students' lax attitudes, disruptive behavior, and fixation with technology; and maintain their own professional development to meet the demands that a 21st century classroom places on its educators. The ideas in this chapter set the stage for a rationale regarding the implementation of the 5E instructional model for teaching science as it relates to each of these components:

- curriculum and instruction
- classroom environment
- professional development

# Scientific Literacy

Together, curriculum and instructional methods are the "what" and "how" of teaching. Science is an adventure in learning facts, figures, and information about the world around us. In addition to scientific discovery and factual learning, students must also understand scientific processes and be able to identify how science plays a critical role in their everyday lives. Collectively, all the facts and information, along with the understanding of the nature of science, the scientific enterprise, and the role of science in society and personal life, make up the definition of *scientific literacy* (National Science Education Standards 1996). To be literate in science, students need to know facts, but they must also be able to experiment, observe, problem-solve, work collaboratively, and think critically. In other words, students must "do" science.

All science teachers should agree on the ideas supporting the concept of scientific literacy. The National Science Education Standards define this as "the knowledge and understanding of scientific concepts and processes required for personal decision making, participation in civic and cultural affairs, and economic productivity" (1996, 22). The characteristics of a scientifically literate citizen include but are not limited to the following:

- Asks and answers questions that have been born out of sheer curiosity about everyday events

- Describes, explains, and predicts naturally occurring events

- Reads and understands articles about scientific topics

- Competently participates in social conversations about scientific results

- Justifies his or her position related to scientific issues that affect society on a local, national, and global level

- Argues scientific conclusions based on evidence

- Uses appropriate terminology

The National Science Education Standards remind us, too, that scientific literacy has varying degrees and forms. Most students come to school with natural curiosity. Teachers discuss and teach about concepts to satisfy students' curiosity and to develop their scientific literacy. However, many students graduate not having completely solidified their scientific literacy. Just as avid readers continue to develop their breadth and depth of reading skills, many people continue to expand and deepen their understanding of science continuously over time. And with the emphasis on STEM (Science, Technology, Engineering, and Mathematics) careers as one option for students' future accomplishments (National Science Board 2007), many will likely graduate from school only to continue on their quest for answers in the workforce.

## What Is the 5E Model of Instruction?

The 5E model of science instruction provides the structure for teachers to meet the demands of today's science standards (both quantitative and qualitative). It engages students' thinking, then allows for explorative discovery and factual learning to deepen students' understanding of content matter. Students learn that one scientific question leads to another, which may lead to several more. Students have the opportunity to become critical thinkers and continue their learning of topics of interest as time passes.

The 5E model is a method of teaching science to produce scientifically literate students. Because it is a pedagogical approach to teaching science, it provides a framework for teachers around which to develop students' understanding of scientific ideas and concepts (content). However, the 5E model of instruction does not support any one program or any one set of material resources. Nor does it define scientific inquiry. This method is flexible and can be used with many different types of instructional resources, programs, and materials that teachers may already have. The 5Es are as follows:

1. Engage

2. Explore

3. Explain

4. Elaborate or Extend

5. Evaluate

As previously discussed, curriculum and instruction refer to the "what" and "how" of teaching. The standards concretely provide educators with the "what." The 5E model structures the "how."

## The 5Es Promote an Engaging Classroom Environment

Classrooms are filled with students who find science boring, unengaging, or otherwise useless. Disruptive behavior, poor attitudes, and downright apathy are not new challenges facing today's teachers. However, today's teachers must accomplish more with seemingly less in the same amount of time as their yesteryear counterparts.

In many science classrooms, lectures and textbooks have historically been part of the instructional approach of many teachers. Unfortunately, when used exclusively, these approaches lead to ritually compliant students, students who sit quietly listening to what may seem to be an endless barrage of uninteresting facts, information, and unknown scientific terms. Students have little commitment to the topic or task, and they devote little attention to what they should know and be able to do. Although lecture certainly has its place in the classroom, today's students deserve to be active learners, engaged and involved in the topic presented to them. An engaging classroom is one where students are both committed and attentive to the task. They value the work they are doing, and they find meaning in the outcome (Schlechty Center 2009). Figure 1.1 illustrates this point. The 5E instructional model allows teachers to assign tasks with both meaning and value to students. Yet, for those who hold their lectures and textbook learning at the heart of their instructional delivery, the 5E model allows room for this, too. It is a win-win for students and teachers.

**Figure 1.1** A Relationship of Attention to Engagement

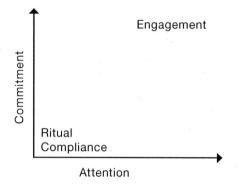

According to Robert Marzano and Debra Pickering (2011), students process several ideas when they are deciding whether they will commit to a topic. First, they consider their emotional investment in the topic. They unknowingly ask themselves, "How do I feel about this?" Second, students consider their interest in the topic. For obvious reasons, if students are uninterested, they are likely to be less engaged. They also decide whether a topic is important to them. The more often a teacher can demonstrate to students that their learning has value to their own personal lives, the greater importance a topic has to students, and the greater their engagement with the learning process. Finally, students consider whether their own skills leave them capable of learning the information or performing a learning task. Students who feel empowered and confident are likely compelled to learn new information. Students who feel incompetent or who lack confidence are likely to disengage from a task. When teachers follow the 5E instructional model—Engage, Explore, Explain, Elaborate or Extend, Evaluate—to teach students about scientific concepts, they can consciously attend to each aspect of the students' perceptions identified by Marzano and Pickering.

The first phase of the 5Es is Engage. When a teacher kicks off a lesson with an activity that boosts the level of classroom energy, students' emotions are heightened, and they become enthused and interested in the topic. By continuing with hands-on activities throughout the rest of the Es, students maintain their level of interest and feel successful and accomplished. Teachers can pull in relevant current events at any phase of the 5E learning cycle, leading students to appreciate the importance of what it is they are learning. Engaging instructional techniques, such as games, controversy, cognitively complex tasks, unusual information, and effective questioning strategies, are elaborated upon in later chapters. Suffice it to say, implementing the 5E instructional model is inherently engaging for students.

When a teacher follows the 5E model, students are free to become active participants in their learning experiences. This is done by way of the careful fluidity of each phase of the instructional model when the teacher is activating prior knowledge, providing time for the exploration of concepts, integrating reading and writing through the explanation of ideas, extending current learning experiences, and evaluating students to provide evidence of learning. These actions collectively formulate critical thinkers in the classroom. Each phase helps students become immersed in science content. This immersion creates exposure to content through varied learning experiences that address all student learning styles and support a variety of teaching styles.

An increased level of engagement leads to other positive classroom environment changes. The use of the 5E model fosters an environment that promotes a positive culture about inquiry. This helps students become problem-solving individuals. Teachers and students become sources of information in the science classroom. The National Science Education Standards (1996) remind us that "[i]nquiry into authentic questions generated from student experiences is the central strategy for teaching science" (31). Through the 5Es, students are able to interact in investigative tasks about concepts and contribute in their classroom environment. This can give students increased feelings of satisfaction and worthiness within the culture and climate of their class, leading to fewer negative outbursts and more productive work. In addition, this model helps students see themselves as scientists rather than spectators in the classroom. In the 5E model, students actively experience learning with an investigative approach as well as a teacher-directed approach. Students' perceptions of science and their role as learners are positively affected. They are allowed and even encouraged to explore, question, and investigate the world around them, which is a natural-born instinct.

# The 5Es Grounded in Constructivist Theory

Constructivism is a learning strategy that builds on students' prior knowledge, ideas, and skill sets. Educators who follow constructivist theory do so following the work of Jerome Bruner. His work was influenced by other theorists, including Lev Vygotsky and Jean Piaget. According to constructivist theory, students formulate new ideas by bridging prior knowledge with new information. Thus, scaffolding takes place. According to Bransford, Brown, and Cocking (2000), scaffolding involves several activities and tasks, such as the following:

- Creating student interest in the task

- Analyzing student needs for current academic task and managing components of the process

- Maintaining the overall objective of the task for the duration of the task

- Evaluating student assessments and products to determine content mastery

- Controlling frustration levels and risk in problem solving

- Providing exemplars/rubrics for student expectations

Clearly, the 5E instructional model abides by constructivist theory. Teachers who utilize the 5E approach in the science classroom tap into students' prior knowledge during the Engage phase. They generate thoughtful inquiry through problem solving, scientific investigations, simulations, or other hands-on activities during the Explore phase. Students' exploration is monitored and guided through inquiry, which allows the teacher to steer students in the right direction without giving them the answer. This way, students synthesize their own learning through problem solving and critical thinking. Information provided during the Explain phase supports students' developing understanding of scientific facts, concepts, and ideas. Students are then offered additional opportunities to build their understanding through Elaboration. Through thoughtful assessment strategies (Evaluation), teachers can help students realize what they have learned and what they have yet to learn, looping back to the Engage phase, when necessary. Ideally, students begin to see learning as a continuous process. The cyclical nature of the 5E model is illustrated in Figure 1.2.

**Figure 1.2** 5E Instructional Model for Science

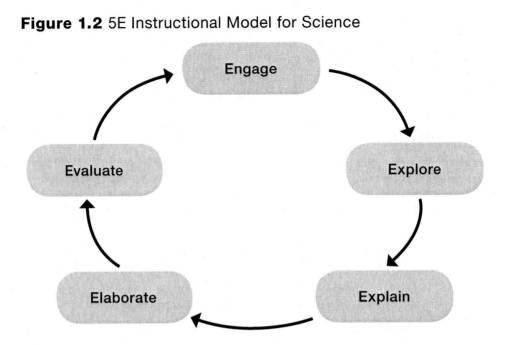

During the Engage and Explore phases, students participate in hands-on or critical-thinking experiences and activities before delving into the scientific content that is introduced during the Explain phase. This is sometimes referred to as the Activity-Before-Content Model, or the ABC Model, which is grounded in constructivist theory. Because of this, the 5E model is inherently based on the constructivist theory of learning, too. Teachers who follow constructivist theory believe that students build, or construct, new learning on top of ideas they already have. Oftentimes, hands-on activities are the mainstay of a constructivist classroom. Generally, these constructivist teachers help students make connections between facts in order to foster new understanding of concepts. These teachers organize learning experiences that require students to analyze, interpret, and predict information. This is sometimes accomplished through open-ended questions, and students are encouraged to engage in dialogue with each other to process their ideas. Constructivist theory is not restricted to any one age group, which allows the 5E model to be successful across many grade levels.

The following is an example of a classroom situation where constructivism was the center of the learning process:

| | |
|---|---|
| **Greg** | *Why aren't all science concepts called laws?* |
| **Teacher** | *There are three words commonly misused or misunderstood in science: hypothesis, theory, and law.* (The teacher writes these words on the board.) *Talk with a partner about these three terms. Between the two of you, decide if you know the differences between these words and if you can explain them.* (The teacher lets students discuss these three terms with partners.) |
| **Jeremy** | *I know a hypothesis is an educated guess. We make these when we do experiments in class.* |
| **Teacher** | *Right. Are they only used during experiments?* |
| **Emma** | *I think so. Marie Curie probably made hypotheses about the experiments she did.* |
| **Jamal** | *Maybe, but she had theories about what she was doing, too. So, is a theory like a hypothesis?* |
| **Tyler** | *You can't have a theory without proof.* |
| **Gigi** | *But you can get the proof when you do the experiments.* |

A class discussion continued in this fashion. When needed, the teacher provided a guiding question to steer the conversation back to the main point to define and explain the differences among a *hypothesis*, a *theory*, and a *law*. Eventually, the teacher moved instruction forward and challenged students to find out the difference among these three ideas at home. The teacher promised they would continue to work toward a common understanding of these terms throughout the week.

What does the previous scenario have to do with constructivism and the 5Es? First, a student, not the teacher, posed a question that had meaning to him. Both his interest and commitment levels were high, so he was definitely Engaged with the topic. Secondly, the teacher did not provide a direct response. Instead, the teacher posed a different question, one that allowed students to make connections to and think critically about what they already knew. Students were allowed to collaborate together and build on their prior knowledge to construct new ideas and justify their ideas with partners. There was no "hands-on" activity, per se, but students were encouraged to Explore more information on their own time, and the teacher promised to devote additional class time to this topic. Hopefully, this scenario illustrated how a constructivist classroom compares to a traditional one. As for the 5Es, the teacher had limited time to devote to an organized, structured 5E lesson plan about scientific laws. However, the teacher masterfully engaged students with a rebuttal question and student-led conversation. The discussion left off having students explore the ideas independently. Once they brought back definitions and examples, students could then begin to construct and Explain the meanings of these terms using the information they found. Then, the teacher could have students Elaborate by asking them to define subsequent ideas as they encounter them in science. For example, in order to engage students when learning about the conservation of energy, the teacher might ask, "What do scientists mean when they say that energy cannot be created or destroyed? Would you categorize this idea as a hypothesis, theory, or law? Why?" Since students initiated this learning, they could, after a suitable amount of time, reflect on their own ideas to ensure a true and complete understanding of these three terms.

## Understanding the 5E Instructional Model

Rather than directly profess the nature of the 5E instructional model, this chapter will walk you through each phase—Engage, Explore, Explain, Elaborate or Extend, Evaluate—by asking you to actively experience each phase of the 5Es by participating in activities that mirror each respective phase. The goal is to develop a deeper understanding of the 5E instructional model and its benefits for both teachers and students. So before delving into the "meat" of the content, let us first Engage our thinking. Once the mind is engaged, we can Explore the content, Explain its specific phases or steps, Elaborate or Extend our knowledge, and then Evaluate our own learning.

## Phase 1: Engage

> What adjectives describe the most highly effective science classrooms? Create a list of as many as you can think of.

Before learning about the 5E model, we want to ready our minds for the content. In this instance, a question was posed for thoughtful reflection, which is one strategy for increasing student commitment and attention, the mere definition of engagement (Schlechty Center 2009). Requiring students to respond to questions allows them to make personal connections to the content, pull from their prior experiences, or otherwise take a stand on one side of an issue or another. If all readers of this book were part of a learning community, defined for our purposes as a group of teachers reading and discussing this book collectively or participating in a virtual book study, we could compare our descriptions of an effective classroom and discuss each person's adjectives. As a group, we could then identify the most common descriptions and work to develop our own working idea of an effective science classroom. Although clearly not possible, this type of discussion and analysis of responses would be the ideal scenario. This first phase in becoming fluent in the 5E model marks a starting point along our journey, one we all now have in common, and one which sets a collaborative and collective goal for both the group and each individual participant.

This particular question has no right or wrong answer, although some educators might challenge others' responses. As long as the learning environment remains supportive and respectful, polite banter is welcome in a 5E classroom. When students question each other, teachers know students are thinking critically (without being critical). Other questions to engage include asking philosophical questions *(Are renewable resources the answer to our energy needs?)*, opinion questions *(Should Pluto be readmitted as a planet?)*, or investigative questions *(What happens when cold and warm air masses meet?)*. In all cases, these questions elicit responses that require some thought. They require students to pull from their own personal experiences or background knowledge in order to assert their own ideas or to make reasonable predictions.

The first question, "What adjectives describe the most highly effective science classrooms?," compels you to brainstorm a list of adjectives. It is open-ended. It requires little to no prior knowledge, drawing only on your experiences. Other effective activities to engage are listed and described in Figure 1.3. However, this list could be extended by including any ideas related to a specific unit of study that wakes up students' minds and requires just a few minutes of class time.

**Figure 1.3** Sample Engage Activities

| Engage Activity | Description | Examples |
|---|---|---|
| Display a picture, primary source, or other visual | Visuals may include data tables, charts, images, or illustrations of cycles. | Pictures may include a Rube Goldberg machine or an aircraft breaking the sound barrier. A primary source might be an example of a Rube Goldberg apparatus. |
| Show a short video clip (2–5 minutes) | Short videos leave students thinking and wanting more information. This is the hook that gets them engaged. | Short videos can be found online on topics, such as weather, erosion, chemical reactions, and cell division. |
| Conduct a simple demonstration | This may be an activity that the teacher does in front of the class or that students conduct individually or with partners. | Examples include banging two pots together to create sound waves that move rice resting on top of a sheet of plastic wrapping or having students manipulate magnets at their seats. |
| Revisit information learned from a previous unit of study | If prior knowledge necessitates learning a new concept, it is worth reviewing. Make it fun! Try a game or game-like activity. | Students might review Newton's three laws of motion before learning about simple machines or the water cycle before learning about climate. |
| Have students reflect on a famous quote or analogy | Other people's words can build our own critical-thinking skills. Use a quote to initiate a class discussion, or have students predict the relationship among ideas in an analogy. | Quote: "In physics, you don't have to go around making trouble for yourself; nature does it for you." (Frank Wilczek) Analogy: "Mitochondria are like engines." |

Oftentimes teachers can use this phase of the 5E model to identify misconceptions students have about a particular topic. Teachers might be alarmed to discover that their students think that all mammals have tails or that there is no difference between rocks and minerals. Once these misconceptions are addressed, teachers can use this information to formulate a direct path of learning to correct these misconceptions. This notion is more fully discussed in Chapter Seven.

## Phase 2: Explore

To follow our application of the 5E model while simultaneously learning its components, the following activity will provide common experiences to act as a frame of reference for the information that is yet to come. For this activity, read the sample lesson plan steps. They relate to a lesson on sound energy. The lesson plan steps are intentionally out of order, according to the 5E model. Match the lesson plan steps with the 5E phase you think that each step demonstrates.

| 5E Phase | Lesson Plan Steps |
|---|---|
| Engage | Read information about how sound is made and how the size and speed of waves affect sounds. |
| Explore | Conduct an activity to create waves by using a string attached to the bottom of a chair leg. Move the string back and forth to create waves that are slow, fast, long, and short. |
| Explain | Look at a picture of a fair. Brainstorm a list of all the things that make sound. |
| Elaborate or Extend | Draw a picture of a playground that includes three or more sounds. Label each sound and describe its sound wave (long, short, high, low, etc.). |
| Evaluate | Revisit the picture of the fair. Sort the sounds from the brainstorming activity into high and low sounds, and loud and soft sounds. Discuss and draw how the sound wave might look for each sound. |

The Explore phase of the 5E model is when students can get their hands dirty and really get to the heart of "doing" science. When students explore science, they conduct scientific experiments or investigations, participate in simulations, or otherwise explore scientific ideas through hands-on activities.

## Phase 3: Explain

The third phase of the 5E instructional model consists of learning facts and information. Here, teachers generally teach students what they need to know. This may be accomplished through many means, such as text or video support, guest speakers, lecture, and so forth. Students may listen or read, take notes, or they may participate in any number of highly effective reading strategies, such as SQ3R, Jigsaw, or Reciprocal Teaching. Here, too, is when students learn essential vocabulary. By now, they should have some experience with at least some terms from the Engage and Explore phases. For example, when creating sound wave models with strings for an Explore activity, the teacher can interject related terms, such as *frequency, pitch,* and *amplitude,* as students discuss the length and height of the waves they create.

To experience the Explain phase yourself, your job here is to read about the history and benefits of using the 5E instructional model for teaching science. Afterward, there is a short written assignment. In the classroom, teachers might use the assignment with students to allow them to reflect on the information they were to have learned.

## The History of the 5E Model

Like all things in education, the 5E instructional model has evolved over time to conceptualize the work of many researchers. In a report prepared for the Office of Science Education National Institutes of Health, Rodger W. Bybee et al. (2006) explain the origin of today's 5E model. Its roots date back to the work of Johann Friedrich Herbart, a German philosopher who influenced American education around the turn of the 20th century. He believed the purpose of education was to develop character. He proposed to accomplish this development through interest and conceptual understanding. John Dewey, who began his career as a science teacher, also had his hand in this modern-day instructional model. He professed the requisite of reflective thinking in the early 1900s. Instructional models subsequently evolved, reflecting each man's ideas about how students learn. In the mid 1980s, the Biological Sciences Curriculum Study (BSCS), supported by grant money from IBM, conducted a design study to develop new elementary-level science and health curricula. The 5E instructional model was one innovation that emerged from BSCS's study. It encapsulates Herbart's ideas of developing interest and conceptual understanding, Dewey's ideas of reflective thinking, and ideas by more contemporary educational researchers by following a set process to first engage the mind, allowing for exploration of concepts.

This model is simply a series of phases that teachers follow to help students reach a deep and thorough understanding of science concepts. Each phase serves a specific purpose. When followed sequentially, the model provides students with a cohesive instructional plan, one that leads "to the learners' formulation of a better understanding of scientific and technological knowledge, attitudes, and skills" (2006, 2). The phases, in sequential order, and the characteristics of each phase may be found in Figure 1.4.

**Figure 1.4** A Summary of the 5Es

| Phase | Characteristics |
|---|---|
| Engage | • Teachers introduce a topic or idea to capture students' interest, curiosity, and attention.<br>• Teachers tap into students' prior knowledge.<br>• Teachers do not seek a "right answer."<br>• Teachers prompt students to reflect on past experiences or current understandings. |
| Explore | • Students conduct hands-on or problem-solving activities or experiments designed to help them explore the topic and make connections to related concepts, often within groups or teams.<br>• Students share common experiences while the teacher acts as a facilitator, providing materials as needed and guiding students' focus. |
| Explain | • Teachers help students observe patterns, analyze results, and/or draw conclusions based on their activities and investigations.<br>• Teachers define relevant vocabulary.<br>• Teachers explicitly instruct a concept, process, or skill to students.<br>• Students demonstrate their understanding of the concept, process, or skill through written work, written reflections, or class discussion. |
| Elaborate or Extend | • Students build on the concepts or ideas they have learned.<br>• Students make connections to other related concepts through new situations. |
| Evaluate | • Students assess their own understanding of the content.<br>• Teachers evaluate students' progress toward mastering the learning objectives.<br>• Evaluation can be formal or informal. |

—Adapted from Bybee et al. (2006)

According to the National Science Teachers Association (2002), students learn science best when:

- instruction builds directly on students' conceptual framework
- they are involved in first-hand exploration and inquiry/process skills are nurtured
- content is organized on the basis of broad conceptual themes common to all science disciplines
- mathematics and communication skills are integral parts of science instruction

The 5E instructional model is an approach to learning that encompasses all of these ideas.

## Phase 4: Elaborate or Extend

After engaging student thinking, allowing students to explore ideas, and teaching them essential facts and information to make sense of a particular scientific concept, teachers may consider their lesson complete. In many disciplines, after teachers teach something, they test students to determine their level of mastery with the content. In a 5E lesson plan, teachers are not quite ready to check for understanding. Instead, the Elaborate or Extend phase of 5E allows students to apply what they learned in a new and novel situation, or to extend ideas to situations outside the science classroom.

Elaborate activities may resemble the activities from any of the preceding three phases. This is also a time for teachers to revisit initial ideas and allow time for students to reshape their thoughts so that students truly own their learning. The following are some ideas that can be utilized effectively during this phase:

- Introduce and discuss related concepts through the use of scientific text, diagrams/visuals, video, Internet, or other media sources.
- Revisit questions that arose during class discussions and have students demonstrate learning in a creative way (e.g., presentation, poster, experimental design, or model).
- Provide additional hands-on learning experiences related to the concept.
- Email or blog with the scientific community regarding student questions or other information related to the concept being studied.

Let's consider the Engage activity at the beginning of this chapter. You were asked to list words to describe a model science classroom. By this point in the book, you have been introduced to the first four phases of a 5E lesson plan. Now, an effective Elaborate activity would be to revisit your list and eliminate words you no longer believe apply and add words you may have omitted. Then, to really get you thinking, prioritize the five most important descriptors by numbering them "1" (most important) to "5" (least important). Ideally, this activity would be completed collaboratively with partners or in small groups; however, given the limitations of how you may be reading this book, you may have to think on your own.

## Phase 5: Evaluate

Throughout each phase of the 5Es explained so far, there have been a myriad of opportunities for formative assessment. These are informal measures teachers use to determine whether students are meeting the learning objectives, often referred to as "checks for understanding." Depending on the results, these checks for understanding allow opportunities for teachers to cycle back through the various phases of the model to ensure that students are learning the necessary objectives thoroughly. This idea is illustrated in Figure 1.5. Only the classroom teacher can make these instructional determinations and modify the instructional plan to better meet the needs of students along their path to becoming scientifically literate. Chapter 6 discusses formative assessment strategies and how teachers may use responsive data to enhance, support, and extend learning for students.

**Figure 1.5** Responsive Nature of the 5E Instructional Model

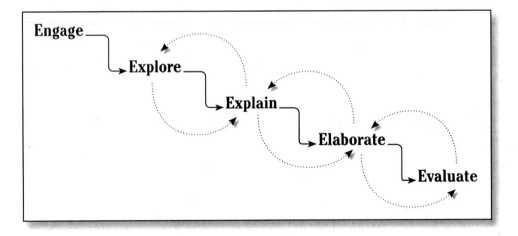

The Evaluate phase of the 5E model is truly designed for summative assessment. It allows for time to assess whether students actually met the intended learning objectives. This summative assessment can take many forms but should ultimately be used to determine students' final understanding of the concept. Some sample assessment ideas are as follows:

- traditional test
- written essay or journal entry
- culminating product
- presentation
- creative project

Now that you have a general understanding of the 5Es, write down four to six terms essential to understanding this instructional model. Then, create a concept map of the information, one that will help you remember the important ideas.

# Benefits of the 5Es for Students

Welcome to the 21st century. This is the age of accountability in education, and not accountability in just reading and mathematics. This includes science, social studies, and writing, too. We want students to know it all. This is the age in which the leaders of the education community (and the community at large) want students to be competent, confident, and productive citizens. What does this mean for teachers? They need to consider how their instruction prepares students with 21st century skills. These skills have been identified by the Partnership for 21st Century Skills (2011).

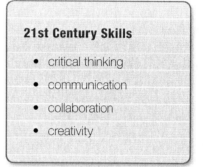

**21st Century Skills**

- critical thinking
- communication
- collaboration
- creativity

Essentially, students need to learn how to think for themselves and use prior knowledge to figure out and master what they don't yet know. The 5E instructional model, because it is well grounded in constructivist theory, provides opportunities for students to practice each of these 21st century skills along their path to becoming scientifically literate.

## Conclusion

The Explore activity presented earlier in this chapter asked you to assign a 5E phase to each of the activities in a lesson plan. Now, during the Evaluate phase of 5E, revisit your initial ideas to confirm your predictions or make changes as needed. Since there is no absolutely correct 5E lesson plan, I choose not to disclose how I would have organized the lesson plan steps to fit a 5E model. Instead, I encourage you to collaborate, compare, and justify your ideas with others.

Additionally, use the Questions for Reflection at the end of this chapter to review the information presented and reflect on your own learning thus far. As with any good scientific inquiry, learning leads to more questions. Now is also an opportunity to record additional questions that you may have.

Today's science teachers are motivating the minds of future inventors, researchers, and scientists. Teachers are, without a doubt, responsible for educating today's children for tomorrow's society. As such, students must do more than simply memorize facts, terms, and information to be considered scientifically literate. The key to learning in the science classroom is helping students make ideas their own. This is best accomplished through varied experiences and a comprehensive, systemic learning cycle. These phases work together as the 5E instructional model to achieve the goals set forth in the Next Generation Science Standards (2012).

## Questions for Reflection

1.  What is the 5E instructional model and what are its phases?

2.  What are the benefits of following the 5E instructional model during science instruction?

3.  How does the 5E instructional model mirror your usual science lessons? How is it different?

4.  Which phase of the 5E instructional model do you believe will be the most challenging for you to implement? Why?

5.  What questions do you still have about the 5E instructional model? After reading subsequent chapters, revisit these questions to see if they have been answered. If they have, place a star beside them and record the page numbers for future reference. If they have not, discuss your question(s) with a trusted colleague or by reflecting on your thoughts in a journal to explain why you think they may not have been addressed.

# Phase 1: Engage

How would Sir Isaac Newton answer the age-old question, "Why did the chicken cross the road?"

During the Engage phase of the 5E instructional model, teachers introduce content to students. This involves gaining students' attention and getting them to commit to the topic. Think of this like a television commercial. If it is too short, our curiosity may be unsatisfied—too long, and we may lose interest. The Engage phase serves as a "commercial" for students' minds. It must be interesting enough to grab their attention in just a short amount of time.

### Effective Engage activities:

- are short in duration
- pique students' interest
- personally involve students
- tap into and assess students' prior knowledge

There is no one right or wrong way to conduct an Engage activity, as long as it accomplishes what its name implies: engage. In particular, time limitations vary from class to class and situation to situation. For example, teachers who take their students on an observation walk to look for evidence of the reflection and refraction of light will likely need more time to conduct this Engage activity than teachers who give their students a magnet and ask them to find three things it is attracted to. Both these examples are engaging activities, but walks typically require more than just a few minutes of class time.

## A Rationale for Engaging Student Thinking

Continuing with the commercial metaphor, think about your favorite commercial. What compels you to watch it? Perhaps something unexpected happens. Perhaps it challenges your own moral compass. Perhaps it includes cute and cuddly animals or even disgusting blobs of mold. Perhaps it is about a popular topic. Perhaps it is just strange. Regardless, it is a favorite and captures your attention each time. Engage activities in the science classroom are like popular commercials on television. They grab students' attention and leave them asking questions, wondering about what is to come, and simply wanting more. Once students are engaged and attentive, they are more likely to commit to the topic, thus elevating their engagement of the entire learning process.

Madeline Hunter, one of the first leaders in instructional best practices, identifies the need to begin every lesson by engaging student thinking (1982). She calls this the *anticipatory set*. Experienced teachers likely wrote many lesson plans as college students that followed Hunter's lesson plan model. Each lesson should begin with this simple step. As evidenced in this chapter, there are many ways to engage student thinking.

**Review Time!**
According to Marzano and Pickering (2011), students ask themselves four questions to determine whether they will engage in a topic:

1. How do I feel?
2. Am I interested?
3. Is this important?
4. Can I do this?

As discussed in Chapter 1, Robert Marzano and Debra Pickering (2011) have identified four essential criteria that must be in place for students to be fully engaged: their emotional investment in the topic, their interest in the topic, whether the topic is important to them, and whether their skills leave them capable of learning the information or performing a learning task.

Regarding difficulty and the level of complexity, engaging student thinking sounds very easy. However, in practice, asking one simple question (a useful Engage activity) can lead to a time-consuming discussion, leaving little time to move on to the remaining 5E phases. Consider this example:

> Mrs. Smith asked a class of fifth-graders, "Should the United States government require all of its citizens to get a flu shot?" She was excited to discover that this seemingly simple question led to a 40 minute debate about the imposing presence of the government in the personal lives of its citizens versus the responsibility of the government to ensure the safety and welfare of all of its citizens.

Mrs. Smith's time restrictions did not meet the intended brevity of the Engage activity. But because the Engage question spurred emotional responses from students and required them to justify their positions, she let the debate run its course and modified the activity that followed. Mrs. Smith wanted students to explore their personal viewpoints. She justified this instructional decision by remembering the purpose of an Engage activity: to jump-start student thinking and hook them into wanting to learn more. This question definitely accomplished its objective.

## Time-Saving Tips

Teachers should set the expectation that their students will discuss a new topic. To learn, students need to interact with each other. As illustrated by Mrs. Smith's class discussion, paired, small-group, or class discussions can sometimes last longer than anticipated. To save time, teachers can have students discuss only one question. They can set time limits (30–90 seconds) for responses and rebuttals. In order to address additional questions that may arise during the discussion, teachers can have a "Parking Lot" chart where students write their questions on sticky notes and post them on the chart for later consideration. Some teachers may feel inclined to answer every question or have a class discussion about everything students want to discuss. This is just not feasible. Teachers should get in the habit of saying, "That's a really great question. Please post it on the 'Parking Lot' chart so we remember to go back to it."

Not all Engage questions need to be answered wholly. When students ask questions, they are sharing their interest in the topic. But one way to maintain their interest is to leave them hanging. Teachers can set students to the task of finding out more about the topic themselves. This way, students begin to answer their own questions. Teachers can reserve class time for students to share their research. A student who is interested enough to independently seek out answers is well on the way to becoming a self-motivated, lifelong learner.

**Tech Tip!**
Instead of physically posting questions to a "Parking Lot" chart, students can virtually post them to a classroom blog or wiki. Setting up a blog or wiki is easy. Visit http://www.blogger.com or http://www.wikispaces.com to get started.

Allowing respectful discussion among students before committing answers to the class requires all students to engage in conversation. Known as the Think-Pair-Share strategy, students first think to themselves and then turn to a partner to discuss their ideas. Then, students share their conversations with the whole class. According to Echevarría, Vogt, and Short (2008), this strategy is especially supportive of struggling students and English language learners (ELLs). The more opportunities a teacher provides for students to listen to and use oral language in a nonthreatening environment, the more confident they will be in their abilities to act as contributing members of the class.

The subsequent sections of this chapter will help teachers develop effective Engage strategies for their classrooms. The strategies are divided into three categories:

1.  strategies to gain students' attention

2.  strategies to personally involve students

3.  strategies to activate prior knowledge

In practice, any one strategy can fit into any one or more categories. For example, teachers can gain students' attention *and* engage them personally by involving them in a class discussion. Regardless of the task, teachers should keep the purpose of the Engage activity in mind: to jump-start student thinking and to get students interested in the learning that is yet to come.

## Strategies to Gain Students' Attention

Generally speaking, anything unexpected, cognitively challenging, emotionally pleasing (or displeasing), or of high interest to students will grab their attention. The following strategies can be used to gain students' attention:

- teacher demonstrations

- pictures and media files

- jokes, facts, and realia

- songs

- stunning vocabulary

- news articles and current events

### Teacher Demonstrations

The element of surprise is always a big hit in the classroom. A lot of science phenomena can seem like "magic" to students, particularly when they see something for the first time. These unexpected incidences have the "Wow!" factor many teachers use to hook students both cognitively and emotionally. Choruses of "Wow!," "Did you see that?," and, "How'd you do that?" are music to a science teacher's ears! Remarkably, these demonstrations do not need to necessarily involve explosives or dangerous chemicals (although these help, too). For example, a teacher can hook students into thinking about force and motion by simply demonstrating the transfer of energy from a moving penny to a stationary penny that pushes a third penny off the edge of a table. Students will watch with skepticism and a twinge of anticipation to see if their teacher really can move a penny without touching it. This simple demonstration grabs students' attention, thus heightening their level of commitment to the topic of force and motion.

Teacher demonstrations might take the form of a controlled chemical reaction or a simple presentation. Because of the nature of these demonstrations, these types of attention grabbers generally center around physical science concepts. Teachers should not overlook the responses they can gain from their students by smashing a rock into pieces or bringing in a live worm and touching it to see how it reacts. Any simple task a teacher can conduct in front of students to gain their attention will generate interest in and engagement with the new topic.

## Pictures and Media Files

Nothing can replace the "Wow!" factor of an in-class demonstration. However, with technological advancements, teachers can grab students' attention through pictures and audio-visual media on virtually any topic, especially those that can't be recreated in the classroom. For instance, teachers might compare pictures of a natural rock formation from 100 years ago and today so that students may observe the effects of weathering and erosion over time. Teachers may also show an online video of replicating cells. Both of these visuals are easily accessed by conducting web searches. Students can experience these natural phenomena through technology when they cannot experience them first-hand in the classroom. They say pictures are worth a thousand words. How wonderful to have students use a thousand words to discuss science! For a short list of reliable and student-friendly multimedia sites, see Appendix A.

## Jokes, Facts, and Realia

Teachers can hook students with a funny joke. The joke at the beginning of this chapter was intentionally asked to get your attention. If being funny is a natural inclination, teachers can use any search engine to find the perfect witticism to kick off any science topic.

But this chapter may have just as easily started with a surprising fact. Did you know, for example, that the world's oldest known living tree resides in Sweden and has been growing for over 9,550 years? This would make a great opening to a unit about the life cycle of plants. The world is full of fast facts, brilliant quotes, and strange objects. Any one of these would generate interest in a particular topic. This strategy addresses Robert Marzano and Debra Pickering's (2011) first criterion for student engagement: how students feel about a topic. Any activity that evokes an emotional response will increase students' engagement in a topic.

Visuals and realia also have a powerful effect on students' curiosity. Well-selected realia evoke feelings of wonder and interest, regardless of the subject matter at hand. To get students interested in the power of electricity, for example, a teacher might show students a fulgurite. Or students could observe spiders using prepared specimens and hand lenses to begin a unit on arachnids. A teacher might begin a study of the moon, Earth, and sun by passing around meteorite samples. If students can see it, touch it, or observe it, it will likely gain their attention and engage their interest.

## Songs

Many teachers may not fancy themselves musicians. However, music and rhythm can be powerful learning tools for many students and can provide them with multimodal learning opportunities, especially when combined with creative movement or visuals (Crowther 2012). Consider playing a song and displaying the lyrics for students. Or make up your own lyrics to a familiar tune and have the class sing it together. Internet resources can provide additional support for background music and lyrics needed to gain students' attention through song. A web search for "science songs" can help locate great general resources to support this.

> **Caution!**
> Teachers should be sure to preview songs and other multimedia files before presenting them to students.

## Stunning Vocabulary

Science is full of unusual, interesting, and original vocabulary. Teachers can use this to their advantage to jump-start students' thinking about essential vocabulary as an Engage activity. For the word *transit*, for instance, a teacher could Engage students by asking, "What was that called when Venus (or another smaller celestial body) passed across the sun (or another larger celestial body) on June 6, 2012?" A teacher could also provide each small group with an envelope containing the crucial words they will encounter during a unit of study. As an Engage activity, the teacher could ask the groups to sort and record the words based on their prior knowledge. Then, after the unit, students can repeat the activity and re-sort the words, explaining why they made these changes from the onset of the lesson. You as the reader have encountered *fulgurite*, and *meteorite*, too. Just as these words are fun to say, learning what they mean will pique students' interest in the topic to which they relate. Engaging students with vocabulary activities like these may help the acquisition of new terms seem less daunting and more fun.

## News Articles and Current Events

Perhaps a local activity, event, or issue relates to a current science topic. Environmental issues in particular are at the center of heated debates at the national and international level. Although reading-related activities are typically saved for the Explain phase, teachers can Engage students by reading and discussing a short news article about a relevant event or issue. While simultaneously activating prior knowledge, reading short articles about related topics and conducting discussions can be effective in engaging students. What could evoke more emotion in middle school students about to embark on a unit about genetics than reading a short passage about mutations?

Figure 2.1 provides additional ideas that can be used to gain students' attention during the Engage phase of the 5E instructional model.

**Figure 2.1** Additional Ideas to Gain Students' Attention in Engage

| Science Area | Grades K–2 | Strategy |
|---|---|---|
| Life Science | Play sounds of odd, lesser-known animals. | media file |
| Earth Science | Dig a gloved hand into a container of soil. Ask students if you are playing in *soil* or *dirt*. | teacher demonstration |
| Physical Science | Play a song about motion or position words (up and down, in and out, high and low, under and over, etc.). | song |

| Science Area | Grades 3–5 | Strategy |
|---|---|---|
| Life Science | Show a time-elapsed video of a growing plant. | media file |
| Earth Science | Melt a chocolate bar using a heat lamp to demonstrate radiant energy. | teacher demonstration |
| Physical Science | Show a loaf of bread. Have students discuss and predict what a loaf of bread has to do with the Law of Conservation of Mass. | realia |

| Science Area | Grades 6–12 | Strategy |
|---|---|---|
| Life Science | Show pictures of various ecosystems. Use the pictures to introduce and define this vocabulary term. | stunning vocabulary |
| Earth Science | Read a short article summarizing scientists' predictions about what they think would happen if an asteroid were to hit the moon. | news articles and current events |
| Physical Science | Have students compare and contrast physical models of various atoms. | realia |

# Strategies to Personally Involve Students

One hallmark of engagement is activity. People are typically engaged when they are personally involved in a meaningful activity. Eric Jensen (2008) reminds us, "Involve, don't tell" (218). Much of a teacher's lesson planning should include activities "to engage learners physically and socially so that they are continually interacting and taking action" (219). Engage activities, by their very nature, commit students to the topic and hold their attention. When a teacher uses hands-on activities to open a unit of study, students' engagement is nearly guaranteed. The following strategies personally engage students physically, socially, or both:

- class discussion

- hands-on activities

- data analysis

- friendly debate

## Class Discussion

A simple way to engage all students, regardless of the focus of the lesson, is to get students involved in an initial discussion. This may be accomplished by posing a thoughtful question for students to consider, discuss, and reflect upon: "What is the fastest thing you can think of?" Other discussion starters include setting up a situation and asking, "What if…," or, "What would you do if…," or having students participate in personal reflections by posing a question that causes them to respond with, "I think…," "I remember when…," or, "I wonder…" An example of this type of question might be, "Is one vital organ more important than another?" When posing questions to initiate class discussions, teachers should remember to keep them open-ended. And teachers should always require students to justify their responses.

To ensure involvement during discussions, students can first discuss their ideas with partners or in small groups. After a short time collaborating, the teacher can then call on just a few students to share their thoughts with the whole class. The "Time-Saving Tips" at the beginning of this chapter may help keep the Engage discussion from running too long into class time.

## Hands-On Activities

Another way to ensure active involvement during the Engage phase of the 5E model is to give students something to manipulate. Since the Engage phase is not intended to be lengthy or time consuming, teachers do not want students to become deeply involved in an experiment or investigation. Rather, with limited materials, students might be asked to manipulate them to perform some simple task.

Small tasks, such as making noises with different-sized rubber bands, manipulating differently-shaped magnets, or searching the room for three objects with different weights (light, medium, or heavy) or textures (rough, smooth, or soft), are perfect activities that allow students to make discoveries on their own as a lead-in to the wonderful world of sound, magnetism, or physical properties, respectively.

## Data Analysis

Students will spend time during the Explore phase of the 5E instructional model conducting comprehensive investigations and experiments, some of which may require students to collect and analyze data. However, the analysis of data should not be overlooked as a viable and resourceful activity during the Engage phase, too. Starting a lesson by displaying a chart, table, or other visual representation of data allows students to make predictions about the information they will soon learn. For example, students might look at different population maps of a specific animal species and discuss in small groups how the climate of the region affects the populations of the species. This would give students an opportunity to think about how animals and their habitats are related before they learn about it in depth.

## Friendly Debate

Teachers may agree that students can be naturally contrary and/or argumentative. In formal debates, students should conduct relevant research to support their positions and refute the opposing position. But for the purposes of an Engage activity, the research component should not be part of the debate. Instead, students are given a chance to state their position based on prior knowledge and justify their thinking.

Making a proposition that requires students to take a stance on a controversial topic can lead to passionate thoughts, thoughts that students are encouraged to express freely (but respectfully) in the classroom. The following are the steps to successfully moderating an Engage debate:

1. **Pose a question.** For example, to introduce the topic of alternative energy sources, a teacher might ask, "Should the government invest money in alternative energy?" Or to kick off a unit on health and nutrition, a teacher might ask, "Do the benefits of eating a vegetarian diet outweigh the dangers?"

2. **Poll the class.** Depending on the culture and climate of the class, teachers may wish to have students express their opinions by anonymously voting. Other teachers who believe one student's opinion will not sway another's may have students vote with a show of hands. To increase student engagement, teachers can polarize the room by designating different locations in the classroom as belonging to different opinions and responses. Students then move to the location that supports their personal responses.

3. **Have students with opposing views justify their reasoning for moving to a certain location in the classroom.** This is an opportunity for teachers to record students' initial ideas on a chart or electronic document. A two-column chart (see Appendix C) is a useful graphic organizer for debates. Once recorded, students' initial responses may be referenced following the conclusion of the lesson or unit. This acts as a meaningful review of the lesson content, leading up to formal assessment. It also provides a structure for notes should the teacher have students participate in a more formal debate as a lesson extension.

4. **At the end of the lesson or unit, students can self-reflect.** Did their initial opinions change as a result of learning additional facts and information?

Figure 2.2 provides additional ideas that can be used to personally involve students during the Engage phase of the 5E instructional model.

**Figure 2.2** Additional Ideas to Personally Involve Students in Engage

| Science Area | Grades K–2 | Strategy |
|---|---|---|
| Life Science | Make finger puppets for animals students will learn about. | hands-on activity |
| Earth Science | Have students record the weather today. Then, have them predict what the weather will be like tomorrow. The next day, check students' predictions and mark the observed weather on the calendar. Continue to make predictions throughout the week. | data analysis |
| Physical Science | Go on a five-senses walk. | hands-on activity |

| Science Area | Grades 3–5 | Strategy |
|---|---|---|
| Life Science | Ask students, "What is the strangest animal you have ever seen?" Brainstorm a class list. | class discussion |
| Earth Science | Provide students with yellow and blue clay. Ask students to construct models of Earth and the sun, showing the relative size of each. | hands-on activity |
| Physical Science | Have students straighten out a paperclip and try to spin it when a partner pinches it between the thumb and index finger. Then, have them bend each end of the clip in opposite directions and try to spin it again. Ask students which task was easier, and why. | hands-on activity |

| Science Area | Grades 6–12 | Strategy |
|---|---|---|
| Life Science | Hold a class debate around the question, "Should lakes be closed off from rivers to keep nonnative fish from invading the lakes?" | friendly debate |
| Earth Science | Go on a nature walk outside the building. Have students look for and record evidence of weathering and erosion. | hands-on activity |
| Physical Science | Have students work with partners to drop a penny and a rock (or a sponge and a bar of soap, etc.) from the same height at the same time. Students should try to determine which will hit the ground first. | hands-on activity |

# Strategies to Activate Prior Knowledge

Robert Marzano (2004) reports that background knowledge is "one of the strongest indicators of how well students will learn new information relative to the content" (1). He identifies the fact that students must experience an interaction between stored information and past experiences to build background knowledge. The best way to accomplish this is to "provide academically enriching experiences" (14). These are activities that "increase the variety and depth of out-of-class experiences."

Engage activities tap into background knowledge when they require students to reflect on what they already know and apply it to a new situation. Students access stored information in order to respond to a question, make a prediction, or verbalize ideas. All of the Engage activities in this chapter require students to use prior knowledge to some degree. But Engage activities can also build background knowledge. By personally involving students, teachers provide a common experience from which students may build on what they already know. The activities to activate prior knowledge that follow give teachers more ideas for Engage activities when starting a lesson or a unit of study. They activate prior knowledge and/or provide common experiences for students related to a new topic.

But be cautious when activating background knowledge; oftentimes, students come to class with misunderstandings or limited experiences. Teachers may discover that students have little or no prior experiences from which to make predictions or answer questions. Not all students—both elementary and secondary students—come to school having touched a frog before. If a teacher were to ask students how a frog's skin feels, students may not have the background experience to respond appropriately to the question. Likewise, teachers should not be surprised if even secondary students have misconceptions about the functions of internal organs, among other scientific topics. Not all students may have learned that the lungs and heart work in tandem to deliver oxygen to cells in the body. Teachers who make assumptions about the knowledge their students bring to class may be disappointed or even shocked to realize that what they may consider to be basic understandings are not so basic after all.

The strategies that follow provide opportunities for students to consider their stored information and share a common experience:

- audio-visual and media resources

- freewrite

- brainstorming and concept maps

- KWL

- pretest

## Audio-Visual and Media Resources

In addition to being great attention grabbers, videos and other visual media activate students' background knowledge and can serve to build background knowledge before new learning takes place. For example, most students know what to call the big yellow fireball in the sky (*the sun*). But when beginning a new unit about the benefits of the sun, a teacher might show a short video of the sun to jump-start students' thinking and get them interested in the topic. Then, the teacher may brainstorm with the class, listing who or what uses the sun, and how they use the sun. The video and the brainstorming activity are both part of an Engage activity, but the video puts students in the proper mindset to participate in a discussion about the topic.

Similarly, most secondary students know what gravity is and how it affects Earth and other objects in the solar system. But to move students beyond their current understandings and into the scientific realm of Kepler's Laws and Newton's Law of Universal Gravity, a teacher might show a poster of the solar system and ask students to describe the motion of the planets around the sun. The teacher can go one step further and poll the class as to whether they believe the planets have a circular or elliptical orbit. The students who know the elliptical orbit to be true can attempt to convince their peers during the opening dialogue of this concept, and vice versa. Audio-visual resources are those that communicate information by using sound or images (more so than text). This includes charts, posters, cartoons, realia, field trips, graphics, and pictures. Media typically refers to resources seen and heard on a computer, television, or digital device, or through a projector, such as videos. Many appropriate and engaging media resources for virtually any science topic may be found online. For student-friendly multimedia sites, see Appendix A.

## Freewrite

Without prompting or discussion, what would happen if students were asked to write what they know about polar bears? Primary students may be able to draw pictures of polar bears or give vague descriptions. Intermediate students may be able to give a few more details about these arctic inhabitants. Middle and high school students may be able to record specific details, such as where they live and what they eat. The idea of a freewrite is that students think about what they know and record their knowledge prior to learning new information. Then, during or after the unit, students may reference their freewrites and add new details, modify incorrect details, or cross off unfounded or incorrect facts.

A freewrite can be a valuable pre-learning activity, but teachers must be sure students know at least something about the topic, or else teachers will undoubtedly be faced with a sea of blank papers. To start a unit on cell reproduction, for example, a teacher would likely not ask students to write everything they know about *meiosis* and *mitosis*. Students would likely be encountering these terms for the first time during another phase, so they would have no background knowledge of either of these processes. However, a teacher could ask students to write everything they know about cell division. By middle school, a teacher should have the reasonable assumption that students have heard of *cells* and *division*. So it would stand to reason that even if students had not heard of *cell division*, they could at least make an educated guess about what it is and write about it.

Freewrites should have short time limits. Time limits will depend on students' grade level, prior experiences with a particular topic, and general writing fluency. Teachers using freewrites for the first time may initially tell students that they have two minutes to record their ideas, but then learn that after two minutes, students needed three or four minutes to record their ideas. Then again, teachers who allow three minutes may discover that students finish writing in 30 seconds. Regardless, freewrites engage students' thinking by having them access their prior experiences and understandings and recording what they know as a means to jump-start their involvement in a particular topic.

## Brainstorming and Concept Maps

Brainstorming is related to freewrites in that it requires students to freely list ideas related to a specific topic or category. Students cannot brainstorm without first tapping into their prior knowledge and experiences. Teachers of students too young to write can verbally conduct brainstorming activities as a whole class.

Similarly, students can collaborate to complete concept maps. Concept maps, like brainstorming, are similar to freewrites in that they require students to spend a short amount of time recording their knowledge in an organized format. However, instead of writing narrative summaries or simply listing items, students organize their ideas into a cohesive visual summary. Figure 2.3 shows what this might look like for an elementary unit on matter. When used formatively, before any learning has taken place, the teacher can visually and quickly assess what students know or think they know and use this assessment to better extend, supplement, and correct student understandings.

**Figure 2.3** Pre-Learning Concept Map of Matter

## KWL

Donna Ogle (1986) created the KWL chart as a reading comprehension strategy. Students list what they *Know* in the "K" column, what they *Want* to know in the "W" column, and what they *Learned* in the "L" column. The "L" column is only completed at the end of a lesson or unit of study when students revisit their charts. This common and reliable graphic organizer may be used across all grade levels and content areas by any teacher who wants to identify students' background knowledge and set a purpose for learning more about a topic. One advantage of using KWL charts is that they effectively assess what information students bring to a particular topic *before* planning an entire lesson or unit. Regardless of whether the information is correct, if a student says they *Know* something, it goes on the chart. This is an effective assessment strategy that allows teachers to identify student misconceptions before a lesson begins. A second advantage of using a KWL chart is that it defines a student-centered purpose for learning about this topic. When students record their own questions or make their own predictions, they own their learning outcomes. They are compelled to find out the answers to their questions or discover if their predictions are true. This increases students' attention and commits them to learn more, which is the essence of engagement.

## Pretest

Pretests may sound unengaging. However, when they are conducted in a game-like fashion, or any other fun mode, teachers gain valuable insight into the knowledge their students bring to a topic, and students have fun competing with their peers to win a game or activity. Teachers can use electronic templates or slideshow presentations to build game show-like pretests. When placed in small groups or teams, students collaborate to answer a series of questions. By writing their responses on the board or by electronically responding using a response system, teachers can maintain an environment where groups don't "cheat" by stealing another group's responses.

Any time students think through questions and discussions, they are tapping into and building on their background knowledge. Figure 2.4 provides additional ideas that can be used to activate students' background knowledge during the Engage phase of the 5E instructional model.

**Figure 2.4** Additional Ideas to Activate Background Knowledge

| Science Area | Grades K–2 | Strategy |
|---|---|---|
| Life Science | Display pictures of different plants. Ask students to list what they all have in common. | audio-visual resources |
| Earth Science | Have students make concept maps about the contents of soil. | concept maps |
| Physical Science | Show a video of an astronaut walking in space. Have students compare the movement in space against the movement on Earth. | audio-visual resources |

| Science Area | Grades 3–5 | Strategy |
|---|---|---|
| Life Science | Ask students, "Could a wild cactus survive in North Dakota?" Have students brainstorm and justify their predictions. | brainstorming |
| Earth Science | Have students collaborate in small groups to make concept maps that show what they know about rocks. | concept maps |
| Physical Science | Have students listen to different instruments (wind, string, percussion), making sounds with different pitches. Have students share what they think causes the pitch to change. | audio-visual resources |

| Science Area | Grades 6–12 | Strategy |
|---|---|---|
| Life Science | Write and conduct a short True/False quiz about ecosystems. | pretest |
| Earth Science | Show a short video of warm and cold air masses colliding. | media resources |
| Physical Science | Show a picture of a radio, a microwave, and a flashlight. Ask students what all of these objects have in common. | audio-visual resources |

# Using Engage Activities to Identify Misconceptions

Identifying misconceptions during the Engage phase of the 5E model has been briefly touched upon in this chapter. Oftentimes, because of their extensive knowledge of and passion for science, science teachers make incorrect assumptions about what students know and understand about scientific concepts. For example, some students may understand gravity as a function unique to Earth. They may not know that *all* matter has gravity. Students also are likely to believe that objects have color, not that color is a function of light interacting with matter. A teacher's role is to ensure that students understand information correctly. To be productive members of a 21st century society, students should graduate from high school and college with as many resolved or dispelled misconceptions as possible. The likelihood of correcting or dispelling all students' misconceptions is nonexistent. Fortunately, society values continued learning opportunities well past formal schooling years. As long as students learn to question, reflect, and analyze appropriately, teachers can be confident that they are sending students into the real world with the tools they will need to identify and correct their own misconceptions as they continue to learn throughout life.

The preceding Engage strategies offer opportunities for teachers to identify students' misconceptions about many topics. The concept map shown in Figure 2.3 allows the teacher to recognize that students know about *evaporation* (although they did not use this term explicitly), but also to recognize that they do not understand this concept fully. With misconceptions identified, teachers can plan lessons and activities that lead students to a more comprehensive understanding of the topic at hand.

## Integrating Writing with Engage

The Common Core State Standards (2012) require students to write throughout the learning process in their content areas, specifically in science. Writing is a natural component of the Engage phase of the 5E model. Since Engage activities are short in duration, students will not have time to write full, detailed pieces. However, they can quickly jot down notes or ideas, or brainstorm lists to reference during later phases of learning.

When a discussion is followed by a writing activity, teachers demonstrate their expectation that all students maintain a certain level of engagement. A short, simple writing activity holds students accountable for their learning and provides the teacher with feedback about students' levels of knowledge. This may be a written prediction, a brainstorming list, a concept map or graphic organizer, or even a one-word response. In the primary grades, brainstorming may be conducted orally with partners, then shared as a class. Teachers should encourage active participation and model how to record ideas in writing so that students will eventually be able to competently, efficiently, and independently attend to this task. As students become more proficient writers, they may begin to undertake responding in writing themselves. One method to support students new to brainstorming is by writing the central question on an anchor chart and having students record their responses on sticky notes and post them on the chart. Students should discuss their ideas with partners before writing out their ideas on the sticky notes (see Figure 2.5). This strategy allows every student to be an active writer. Since the sticky notes have limited space, students do not feel threatened or overwhelmed by the task. By allowing students to discuss their ideas before committing them to paper, you give both struggling students and English language learners the support they need to be active participants in class. In Figure 2.5, the kindergarten class brainstormed a list together, which the teacher recorded on the anchor chart. Then, each student discussed their ideas with a partner and completed his or her own sticky note to add to the chart.

**Figure 2.5** Kindergarten Brainstorming List Sample

Writing in science during the Engage phase should not be intimidating or time-consuming. The ideas in this chapter included suggestions for short, simple writing activities as a natural part of the learning process. These suggestions are summarized in Figure 2.6.

**Figure 2.6** Writing Suggestions for Engage

| Engage Activity | Writing Suggestion |
|---|---|
| brainstorming | write lists, record ideas on sticky notes, write or draw concept maps |
| teacher demonstration | write predictions, record "I think..." or "I wonder..." statements |
| picture observations | record observations, make inferences, write predictions |
| videos or media files | write a follow-up question, summarize |

# Conclusion

To answer the joke at the beginning of the chapter, Newton once said, "Chickens at rest tend to stay at rest. Chickens in motion tend to cross roads" (Morin 2008).

Hopefully, the joke intrigued you and kept you reading further. The reader's reward is finding the punch line here at the end of this chapter.

All joking aside, however, the Engage phase of a 5E instructional model is the key to gaining students' attention in what might otherwise seem to be a boring or uninteresting topic. In this phase, teachers can jump-start students' thinking or simply grab their attention, leaving them wanting to learn more.

Engage activities are short and quick. They provide much insight into students' thinking, yet they take mere moments of class time.

The Engage phase provides an opportunity for teachers to find out what students know and don't know by tapping into their prior knowledge and having them share their understandings before in-depth learning begins. This is a valuable instructional step for all teachers. It helps them identify any misconceptions students may have with the content. Once these misconceptions are identified, teachers may modify the rest of their instructional plan to correct these misconceptions and help students reach a deeper, more thorough understanding of the subject matter.

## Questions for Reflection

1. What are the characteristics of effective Engage activities?

2. Complete this analogy: Engage activities are to the 5E instructional model as _____ are to _____.

3. How might you introduce a new science concept using an Engage activity? Explain what you hope to accomplish with this particular activity. How might this activity help you identify misconceptions that your students may have?

# Chapter 3

# Phase 2: Explore

*"I hear and I forget. I see and I remember. I do and I understand."*
—Confucius, Chinese philosopher (551 B.C.–479 B.C.)

What does it mean to explore something? According to Merriam Webster, to *explore* means "to investigate, study, or analyze; look into; to become familiar with by testing or experimenting; to travel over (new territory) for adventure or discovery; and to examine, especially for diagnostic purposes." In science, exploration leads to a list of scientific process skills, which students must apply to become effective "explorers." As students embark on each new hands-on experience, teachers may likely recognize the skills students use to complete their task. What is important to note here is that an effective Explore activity, usually an experiment or investigation, requires students to simultaneously apply many of these skills.

Scientifically literate students develop the following process skills when they explore the content:

- analyze
- classify
- collaborate
- collect
- communicate
- compare

- describe
- estimate
- evaluate
- follow directions
- graph or chart
- hypothesize

- identify
- infer
- interpret
- measure
- observe
- organize

- predict
- question
- read
- record
- sort
- study

When people learn new information, they tend to do so through experience. Consider an educational finance class, for example. Someone who has experience organizing and setting school budgets prior to taking the class would likely have a relatively firm grasp on the overall finance process. He or she would be able to bring meaning to the otherwise abstract system on numeric codes. Past experience anchors this individual's learning. Someone who has never worked with the codes would lack the concrete experience to bring meaning to them at the outset of the class. He or she can memorize the numbers but would not have a deep understanding of educational finance.

Some teachers are inclined to jump right in and "teach" students the necessary content. But lectures, discussion, and textbook reading do not constitute "doing" science. The Explore phase of the 5E instructional model allows students to create common experiences from which to develop a deeper, more thorough understanding of advanced and abstract scientific concepts. As students complete hands-on investigations, they begin to identify and modify their own conceptual understanding of a topic, process, or skill. As a middle school student, I was certain that water would heat up faster than sand. During a lab activity where I compared the heating rates of equal amounts of water and sand, I realized very quickly that my understanding was incorrect. I learned that my conceptual understanding of how heat affects solids and liquids was incorrect. I was able to make this advancement in understanding because I had discovered it for myself. If my teacher had simply told me this information, or if I had just read about it in a textbook, I likely would not have believed it or remembered it. Therefore, my conceptual understanding would likely have continued to be incorrect. It was the personal exploration I did that led me to new understanding.

Learning science inherently involves learning facts and information. It also involves complex processes and concepts. But students must begin with what may seem like basic foundational knowledge. For example, why should students know the fact that matter is made up of solids, liquids, and gases? In later years, they will likely learn about the water cycle (a process), which in turn leads to understanding weather patterns (a concept), both of which are related to solids, liquids, and gases. To be truly scientifically literate, students need to have opportunities to experience science. Simply reading about ideas and concepts in a book, watching a video, or listening to a lecture will not lead to deep understanding. However, when students touch, manipulate, and investigate objects and materials related to these ideas, they have concrete experiences from which to ground their informational learning.

This chapter shares information related to different types of hands-on experiences: inquiry-based and noninquiry-based. Both types of experiences will help students continue learning during the Explore phase of the 5E model.

# The Definition of "Inquiry"

Students accomplish many tasks during Explore activities, experiments, investigations, and labs. For clarity, these terms are defined in Figure 3.1.

**Figure 3.1** Inquiry Terms Defined

| Term | Definition |
|------|------------|
| activity | something people do; a process or procedure to learn through experience |
| experiment | a test under controlled conditions in order to demonstrate something known or discover something unknown |
| investigation | a measurable study of a scientific question following the steps to the scientific method |
| lab | the act of conducting scientific research |

All of these tasks are similar; they require students to apply process skills. Teachers want to provide varied and novel hands-on learning opportunities so that students can begin to actively construct their own knowledge, thus leading to deeper understanding of subject matter. However, not all inquiry-based hands-on activities are created equal. According to Randy Bell, Lara Smetana, and Ian Binns (2005), "in order for an activity to be inquiry-based, it must start with a scientific question" (31). Skills, such as reading a thermometer, measuring liquid volume, or even assembling a model of the solar system, do not qualify as inquiry-based activities. Students are not, in these instances, centering their work around a scientific question. The Explore phase of the 5E model is less concerned with skill-building or model-building and instead is more concerned with actual investigations that require students to ask, consider, and attempt to answer questions.

"Scientific inquiry," according to the National Science Education Standards (1996) "is at the heart of science and science learning" (15). This is part of what makes the Explore phase so important. When students participate in an inquiry-based activity, they engage in a task by questioning, analyzing data, and thinking critically. They combine knowledge and understanding with reasoning and thinking. And they communicate results and compare results with others. Regardless of whether the teacher poses questions or students devise their own, students are involved in inquiry when they conduct their own analysis and draw their own conclusions. To be true inquiry, students must answer research questions through data analysis (Bell, Smetana, and Binns 2005). Teachers can use experiments, investigations, and labs to support this theory behind the Explore phase of 5E. Although students typically apply skills, such as reading and measuring, to achieve their objective, these skills are not inquiry-based in and of themselves. Instead, they are described as process skills. Likewise, students usually enjoy designing and creating two- or three-dimensional models as products to demonstrate their understanding of scientific knowledge. Such activities include making flowers from baking cups with green paper for the stems and leaves and brown frayed yarn for the roots, or an edible example of an aquifer or plant cell. These activities demonstrate the knowledge students have learned. However, they do not qualify as hands-on, inquiry-based activities and are, therefore, not examples of model Explore activities.

## Levels of Inquiry

Inquiry-based experiences have different levels of complexity. Some inquiry-based activities are quite simple. They demonstrate known outcomes. Students simply create an environment to personally demonstrate the effects. Some inquiries are quite complex. They require students to generate their own questions, create their own experimental designs, and truly discover something unknown. Other inquiry-based activities fall somewhere between these two levels of complexity. Marshall Herron (1971) developed the "Herron Model," shown in Figure 3.2, to illustrate varying levels of complexity with regard to inquiry activities. The complexity levels are organized from the simplest inquiries (0) to the most complex (3).

No one level of inquiry is better than another. However, if the ultimate goal of science education is producing scientifically literate students, then teachers should feel compelled to provide varied inquiry-based activities. This allows for students, over time, to begin to take responsibility for their learning and to apply their analytical and communication skills toward specified outcomes.

**Figure 3.2** Herron's Model of Scientific Inquiry

| Level of Inquiry | | Question Posed by | Procedure Prescribed by | Solution |
|---|---|---|---|---|
| 0 | confirmation/ verification | teacher | teacher | known |
| 1 | structured | teacher | teacher | unknown |
| 2 | guided | teacher | students | unknown |
| 3 | open | students | students | unknown |

## Confirmation/Verification Inquiry

The first level of inquiry is the confirmation/verification level. These types of inquiry-based activities are usually entirely teacher-directed. The teacher provides the question, the teacher dictates the procedures, and the teacher knows the outcome in advance. For example, if a teacher asks what might happen to a rubber ball when it is gently dropped or when it is forcefully thrown to the ground, students will likely have enough background knowledge and experience with rubber balls to know that the harder they are thrown, the higher they bounce. To demonstrate this, either the teacher or the students can try it out. This is an inquiry-based Explore activity since students are relating their hands-on experiences to a scientific question. However, the outcome is known before the activity even begins. The act of gently dropping or forcefully throwing the ball to the ground confirms or verifies students' knowledge.

## Structured Inquiry

The second level of inquiry is a little more complex. It is still mostly teacher-directed, but the outcome is unknown. For example, to keep with the rubber ball activity, the teacher might ask, "How does the surface of the ground affect the height of a bouncing rubber ball?" For this investigation, students would make their predictions in response to a teacher-posed question, then follow the teacher's procedures to discover how a surface (e.g., carpet, concrete, and sand) affects the height of a bouncing ball. Students may have an idea about the outcome, but the activity might lead to more questions and further investigation. Herron identifies this level of inquiry as structured.

## Guided Inquiry

Beyond structured inquiry is guided inquiry. With this level of complexity, students become more actively involved in the process of learning. They know the question they are to answer because it was provided by their teacher. However, they are left to their own devices to design and construct an experiment to test their ideas. They have no idea what the outcome might be. The teacher might ask, "Does the size of a rubber ball affect the reaction force when it strikes the ground?" in order to guide students' predictions. After making their initial predictions, students would set the procedures and acquire the necessary equipment and materials to test their idea. Afterward, a class discussion might uncover missteps in the experimental design or conflicting results. Perhaps one group only measured the diameter of the rubber ball but did not measure its mass. If this were the case, students could not verify their results because they would not know whether force was a function of the diameter or the mass of the ball. This level of inquiry may seem harsh to some teachers. After all, how could teachers intentionally allow students to incorrectly construct an investigation? What is important to remember is that as students develop into functioning 21st century learners, they need to experience missteps so that they can strengthen their analytical skills and become truly independent thinkers. Students whose teachers constantly feed them information or correct their every error never have a chance to analyze their own errors or apply new skills. It's okay to let students explore on their own. There will be time after the investigation to identify missteps, discuss how their design might have been improved, and motivate students to try again with a different approach.

## Open Inquiry

The most student-centered inquiry-based activities are identified as open. This level of inquiry is almost entirely student-directed. The teacher acts as a facilitator, providing only material resources and organizational support. All of the work is completed by students. Since the question originates with students, illustrating an open inquiry-based activity is somewhat difficult. In the classroom, a student might wonder, "Can I make my own rubber ball that out-performs the best one in class?" To investigate this idea, the student would need to research formulas for creating rubber balls, create them, then test his or her products against the rubber balls at school. All the procedures—materials, data collection, data analysis, and summaries—are determined by the student. The teacher might guide research questions by posing other questions students may not have considered, but in essence, students conduct the inquiry on their own.

# Experimental Design

In science, students conduct investigations for one of three purposes:

1. to demonstrate a known outcome

2. to test a hypothesis

3. to discover an unknown outcome

Regardless of the purpose of the investigation or the level of teacher support, true inquiry-based science typically follows a series of steps known collectively as the *scientific method*. These steps are shown on the following page.

## Steps to the Scientific Method

1. Ask a testable question.

2. Conduct research and form a hypothesis.

3. Test the hypothesis and gather and organize data.

4. Analyze and interpret data.

5. Draw a conclusion, summarize, and communicate the results.

## Ask a Testable Question

A true inquiry-based activity focuses around a question. Curious students ask questions all the time. *Why is the sky blue? When will the rain stop? How much air can I blow into this balloon before it pops?* These are all relevant and meaningful questions students ask. Although teachers encourage questioning and curiosity, not all questions meet the criteria to be considered inquiry-based. In order for a question to be tested using inquiry-based methods, it must:

- be testable
- be measurable
- be observable
- lead to further inquiry

The first question *(Why is the sky blue?)* is very common. However, it is not testable in the classroom setting. Understanding how sunlight interacts with Earth's atmosphere is a high-level and abstract concept. While students can conduct experiments to learn about reflection, refraction, and absorption of light to understand color better, the answer to this question comes from knowledge and understanding of light concepts. Since this particular question is not testable, students cannot conduct an inquiry-based activity to answer it. This does not mean that teachers don't want students asking this question. It just means that as far as the Explore phase of the 5E model goes, it is not an appropriate question.

The second question *(When will the rain stop?)* is a cause-effect question. To answer this question, students need a general understanding of weather patterns, the water cycle, and seasonal trends in the area. This question is also not testable. Teachers could have students conduct inquiry-based activities related to hot and cold air masses, wind patterns, and high and low pressure systems. But again, this particular question itself is not testable. Students could, given the right tools and background knowledge, use up-to-date weather maps and weather instruments to make a prediction about the rain, but they could not test their prediction in the classroom.

The third question *(How much air can I blow into this balloon before it pops?)* is testable. Testable questions lead to experiments. If a student wants to test the limits of the latex, he or she can blow until the balloon reaches its maximum limit and pops. But then what? Well, the student tries again with another balloon to see whether it yields the same results. The difficulty with this particular question, however, is that quantifying these limits could pose a challenge. How would a student determine whether one balloon holds more air than another? He or she could count the number of blows. But the amount of air could vary from blow to blow and from balloon to balloon. Therefore, this measuring plan may not yield accurate or valid results. The student could devise a means of measuring the width of the balloon as it nears its limit, but this would have to be done the very moment before the balloon pops. Perhaps the student could wrap a string around the diameter of the balloon and measure the string to gauge the circumference of the balloon. This, too, might not be the most accurate measuring system. The string might slide, or the student holding it might leave more or less slack in the string from balloon to balloon.

The point is, when conducting an inquiry-based activity, teachers want to show students how to ask questions that can be tested, measured, and analyzed based on their evidence. Sometimes the questions are connected to an overall scientific concept, but they do not match exactly. For example, a teacher may want students to explore the relationship between arm span and height as part of a unit on the human body. Asking, "Is there a relationship between arm span and height?" is simply a *Yes* or *No* question that really does not lead students to further inquiry about the topic. However, this question would work for an Engage discussion where students think about this idea and discuss whether it is true. Asking, "What is the relationship between arm span and height?" inherently leads students to the conclusion that there is a relationship and that they need to find out what it is. To test this question, the teacher might ask, for the purposes of conducting an inquiry-based activity, "What is longer: people's arm spans, heights, or are they exactly the same?" This question is testable. Students can measure their arm spans and heights, compile their data, and analyze the results. This will inevitably lead to discovering that, yes, there is a relationship. Teachers can also use this opportunity to teach students about mathematical correlations. Online websites allow the class to chart their data and identify the correlation. They can even use computer-generated formulas to predict people's heights when their arm span length is known (or vice versa), which brings in the application of algebraic formulas. This inquiry also leads to other questions, such as, "Is the correlation among children stronger than the correlation among adults?," or, "Is there a correlation between people's arm spans and their height?" All the features of this question (i.e., that it is testable, measurable, observable, and leads to further inquiry) are evidence that students are exploring using inquiry.

Asking good inquiry-based questions is the mainstay of experimental design. So what are good inquiry-based questions teachers and students can ask? Figure 3.3 provides true inquiry-based questions students can investigate during the Explore phase of the 5E instructional model.

## Figure 3.3 Examples of Inquiry-Based Questions

| Science Area | Grades K–2 |
|---|---|
| Life Science | How does a worm react to light? |
| Earth Science | What causes day and night? |
| Physical Science | What happens on the outside of a cup when ice is added to water? |

| Science Area | Grades 3–5 |
|---|---|
| Life Science | How does gravity affect plant growth? |
| Earth Science | How does warm salt water interact with cold tap water? |
| Physical Science | How strong is the force of water? |

| Science Area | Grades 6–12 |
|---|---|
| Life Science | How quickly do worms decompose vegetables? |
| Earth Science | How does the gulf stream affect ocean currents? |
| Physical Science | How do substances such as salt and sugar affect the density of liquids? |

## Conduct Research and Form a Hypothesis

The second step to the scientific method is conducting research to become further informed about the question being tested. Research during this phase allows students to develop an informed hypothesis. Research usually occurs when students are completing an open-level inquiry. When students are left to their own devices to answer their own question, they need background research to make reasonable predictions. However, during the Explore phase, most investigations are teacher-directed to some degree. During this phase, teachers may skip the research step of the scientific method and instead use it during the next phase—Explain. Research during the Explore phase usually amounts to students discussing what they already know and applying their background knowledge, observations, and past experiences to formulate a reasonable hypothesis. Using the same arm span/height investigation, students could simply hold out their arms, put their hand on their heads, teeter sideways, or observe each other's arm spans to make predictions. The answer is limited: arm span is longer, height is longer, or they are the same. Once students form a hypothesis, they are ready to get to work.

## Test the Hypothesis and Gather and Organize Data

During this stage of scientific inquiry, students can design and/or follow a procedure to test their prediction. Since inquiry-based questions are testable, students will need to collect data. This may take the form of a table, list, or chart. Students who are learning how to collect data may benefit from using a teacher-created organizer to record their data. Figure 3.4 shows what a table might look like for the arm span/height investigation. The teacher could provide the chart for the class and students could record the data in the chart once measurements are taken.

**Figure 3.4** Sample Arm Span/Height Data Collection Chart

| Student | Arm Span (cm) | Height (cm) |
|---|---|---|
| Student 1 | 132 | 132 |
| Student 2 | 132 | 137 |
| Student 3 | 128 | 138 |
| Student 4 | 127 | 139 |
| Student 5 | 142 | 142 |
| Student 6 | 142 | 146 |
| Student 7 | 150 | 152 |
| Student 8 | 158 | 160 |
| Student 9 | 174 | 170 |
| Student 10 | 178 | 174 |
| Student 11 | 140 | 140 |
| Student 12 | 141 | 140 |
| Student 13 | 166 | 167 |
| Student 14 | 150 | 154 |
| Average | 147 | 149 |

Having class data allows students to calculate the average arm span lengths and compare it to the average height of students. The class could also tally the number of students whose arm spans were longer than their heights, the number of students whose heights were longer than their arm spans, and whose were equal. This data could also be easily transferred to an electronic scatter plot to determine whether there was a relationship and what this might be.

Figure 3.5 illustrates how the class data for the arm span/height investigation might look in a scatter plot.

**Figure 3.5** Sample Scatter Plot Graph Comparing Student Arm Lengths to Heights

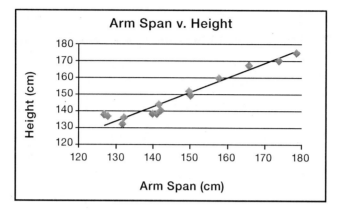

## Analyze and Interpret Data

Without analysis, the data students collect and display are just data. Without careful review, analysis, and interpretation, they are just numbers on a page. Having data organized in a visual display (e.g., a chart or graph) helps students begin to analyze and interpret the data in order to create meaning and draw conclusions. At this point, students should discuss the data in small groups and record their observations. Then, the teacher can guide students to use specific data to make comparisons. In the case of the arm span/height data, students can compare the actual correlation to a perfect correlation (1.0). The teacher, knowing this may be a new math skill for some students, should make sure that this skill is taught in advance so students can make sense of the data.

Students new to analyzing data may need more direct support from the teacher in the form of guiding questions. These questions might include:

- What facts can I state about the data?

- How does the data compare from person to person, from object to object, or from day to day?

- Did the data show me the results I had expected? Why or why not?

- What surprised me about the data?

- What conclusion, if any, can I draw from the data?

## Draw a Conclusion, Summarize, and Communicate the Results

Finally, students reach the final step of the inquiry process: drawing a conclusion and summarizing findings. The teacher may redirect students to the initial question to remind them of what they were attempting to answer. Then, using their analysis of the data, students may begin to formulate a conclusion. This is a very challenging process for some students. Many students tend to respond in generalizations instead of citing specific details to justify their generalizations. For example, when summarizing the arm span/height data, students may think a one-sentence summary is satisfactory: *My arm span was longer than my height.* Although not incorrect, it omits a great deal of information, such as how the student's personal data compared to the rest of the class, the correlation between arm span and height, and the ability of students to predict height if arm span is known. Students need support in constructing these conclusions and models of what an appropriate conclusion looks like. To do this, model writing a conclusion as a class as a shared writing experience. While writing, discuss the following qualities of a good conclusion:

- It outlines the results of the data.
- It compares individual data to group data (if applicable).
- It explains what the results mean.
- It explains why the results are important.

Once students have practiced writing conclusions as a group, they will likely experience greater individual success when they write conclusions on their own.

# Noninquiry-Based Activities Worth Exploring

Not all questions students and teachers ask necessarily lend themselves to inquiry-based investigation. Some activities are just that: activities. However, that doesn't mean that students and teachers cannot address these questions. Students can participate in noninquiry-based activities throughout their scientific quest to learn about the world around them. For example, to answer the question, "What is the structure of DNA?," students might manipulate helixes on an interactive computer program to complete a strand of DNA. To answer the question, "How do certain rocks compare to other rocks?," students may use measuring tools and observations to compare and categorize rocks. Students might observe and record living organisms in a small patch of grass outside to answer, "What lives outside my window?" Or students may simulate how bones work using pipe cleaners to explore the question, "How do bones and muscles work together?" For any of these questions, students explore science. They are actively involved. They are learning about their world; they are not, however, testing anything. Therefore, students are not truly involved in an inquiry-based activity. Much of the content of this chapter focuses on having students conduct inquiry-based activities during the Explore phase of the 5E model. Although inquiry-based activities are preferred, some noninquiry-based activities are equally meaningful as students begin to learn scientific concepts. It is wholly suitable for teachers to allow students to participate in noninquiry-based activities during the Explore phase of the 5E instructional model.

During those times when inquiry-based activities are not possible, students can participate in noninquiry-based activities. The following are sample noninquiry-based activities specific to the topic at hand:

- "Digging In"
- "Presto-Change-O"
- "Times Change"

## Digging In

"Digging In" definitely has students explore science. They are applying, analyzing, describing, observing, measuring, and comparing process skills. However, they are *not* testing anything, so this hands-on activity does not qualify as an inquiry-based activity. All the same, with the structured guidance and support of the teacher, students share the common experience of observing soil before they read about the contents of soil. This way, all students can concretely anchor their learning by reflecting on this Explore activity. This will help them reach a deeper understanding of the content and will lead them to consider the essential question, "Are soil and dirt the same?"

*Mr. Johns wants his second-graders to investigate soil. The objective is to have students understand what makes up soil. He could have them conduct an inquiry-based activity to answer the question, "Which type of soil is best for growing pansies?" However, this particular inquiry does not require students to necessarily "dig in." He likes the idea of planting pansies, and he might use this as a lesson extension during the Elaborate phase of the 5E model. He decides instead to have students observe and compare different soil samples. He has potting soil, garden soil, peat, and sand from a local home improvement store. He invites students to bring in soil samples from around their homes or community (with parental assistance) to observe and compare as well. He provides each student with a record sheet where they may systematically organize their observations. He has paper towels, toothpicks, tweezers, hand lenses, primary microscopes, and balance scales available for students to conduct their investigations. After students have made and recorded their observations, they collaborate in small groups to discuss their findings. Once the activity is complete, Mr. Johns feels his students have enough experience with soil to understand the content of the lesson—that is, understanding what makes up soil.*

# Presto-Change-O

With "Presto-Change-O" activities, students explore the idea of physical and chemical changes. Mrs. Kemper, for example, has not taught her students about these concepts yet. In accordance with the 5E instructional model, she has first engaged their thinking. The sorting activity students participate in requires them to discuss and reflect on what they already know in order to make predictions. The stations do not begin with an inquiry. Instead, they all relate to the same question posed at the start of the lesson. Students are not researching this question through data analysis. They are simply recording observations before and after an event, either combining ingredients or melting ice. Once students read and learn about physical and chemical changes, they can use this information to correctly identify each outcome.

> *Mrs. Kemper wants her fifth-grade students to explore the idea of physical and chemical changes. She engages students by asking, "What is the difference between a physical and a chemical change?" After students discuss their ideas in small groups, each group receives a list of actions that can be categorized as either physical or chemical changes. Students sort each action into the two categories and record their predictions. Mrs. Kemper realizes that she cannot allow students to conduct too many inquiry-based activities to learn about each type of change; that would simply be too time consuming. Instead, she decides to divide and conquer. She sets up four hands-on stations: two illustrating physical changes and two illustrating chemical changes. She divides the class into four groups. Each group completes one of the hands-on stations. The group takes notes about its station in a matrix, listing the beginning and ending substances (see Figure 3.6). Following, students share their findings with the class. After students read about physical and chemical changes during the Explain phase, they will label each station outcome as an example of a physical or chemical change. They can also verify their predictions from before the investigation.*

**Figure 3.6** Teacher-Created Matrix for Recording Activity Outcomes: Physical vs. Chemical Changes

| Station Number | Before Activity | After Activity | Physical or Chemical Change? |
|---|---|---|---|
| 1. no-bake cookies | | | |
| 2. slime | | | |
| 3. salt water | | | |
| 4. melting ice | | | |

## Times Change

The online simulation Mrs. Pike uses in the following example to explore the idea of natural selection is like a game. As the simulation progresses, it records data and students review the data to draw conclusions about the fictitious species. By having students participate in this online simulation before learning about the theory of evolution in depth, you give them first-hand experience with studying a species, variations of a phenotype, inheritance, survival, reproduction, and natural selection, all critical to understanding the concept of evolution. Then, as students read and learn about the theory of evolution, they have a common experience from which to build their knowledge: the online simulation.

Mrs. Pike's seventh-grade students are learning about the evidence that supports the theory of evolution. She begins the unit by asking, "What species have evolved over time?" Students recorded their ideas on sticky notes, shared their ideas with the class, and posted the sticky notes to an anchor chart. Before the Explain phase, Mrs. Pike wants students to Explore the concept of natural selection that supports the theory of evolution. She finds an online simulation to conduct with the class. Students identify both the rate of change in a species and how these changes affect other members of the species population. After conducting the simulation, students discuss what they have learned so far about natural selection and evolution with partners and then the whole class.

# Classroom Management Tips for Exploring Science

There is no way around it; the Explore environment is busy and can be messy. That is the nature of scientific exploration. Investigations usually require at least one class period in order for teachers to adequately conduct them. That being said, teachers can take steps to minimize time spent on classroom management in order to maximize student learning time. Here are a few tips to consider:

- be ready
- set clear expectations
- limit teacher talk

Being prepared, defining expectations ahead of time, and limiting teacher talk should help maintain a structured, productive, and successful work environment for both the teacher and the students.

## Be Ready

Teachers who effectively plan and prepare reduce wasted time by staying organized. The first tip to having Explore activities run fluidly is being ready. Visualize how each investigation will run. Will students work independently, with partners, or in groups? This will determine the number of supplies the teacher needs to have ready. It is also important to have equipment and materials ready. This includes student activity sheets, recording sheets, writing materials, online resources, equipment needs, and other investigation supplies. If students will be measuring liquids or solids, they will need appropriately sized beakers, graduated cylinders, or other containers. If students will be measuring length, they will need centimeter rulers, meter sticks, or other measuring tools. If students will be making complex calculations, they will need calculators. Teachers should also consider students' past experiences with using science equipment. Students who will use a triple beam balance scale will need to know, understand, and have practice using this type of measuring device before the investigation.

Using equipment and supplies can be confusing, especially when investigations require several items. One way to have lab materials organized is by placing all the required materials on a plastic tray, in an aluminum pan, or in a paper or plastic bag. A teacher who has five lab groups would then prepare five trays, pans, or bags, each including all the investigation supplies. Then, the teacher can easily distribute these materials to each lab group. If this is not feasible, the teacher can assign certain students to hand out certain equipment. Or the teacher can have each lab group decide on one person to be the "gopher." This person would be responsible for collecting the needed equipment or special materials. No one organizational strategy will work for every situation. Leading up to the investigation, teachers should spend time reviewing the classroom expectations to minimize confusion and remind students of their responsibilities with regard to equipment needs during lab time.

## Set Clear Expectations

Any classroom runs best when the teacher and students have clearly identified the expectations. Both students and teachers must adhere to them or face pre-established consequences. Each teacher has his or her own method of establishing expectations and consequences. Whether expectations are student- or teacher-generated, be sure to post the expectations where students can clearly see them. Review the list before every activity. This can save time in the long run. Expectations might differ from class to class, depending on the class and students. As long as students know the expectations and consequences beforehand, they usually adhere to them in order to participate in the activities. The following is a sample list of hands-on learning expectations.

## Exploring Science Expectations

- Collaborate and communicate with each other.
- Think about your work.
- Be respectful of others' ideas.
- Resolve conflicts respectfully.
- Talk quietly.
- Use equipment safely and responsibly.
- Clean up.
- Watch for and respond to teacher signals.

## Limit Teacher Talk

Finally, remember that lab time is student learning time. During lab time, the classroom should be filled more with the buzz of students learning and less with the teacher talking. On Explore days, quickly set students to their task by limiting teacher talk. Review the initial inquiry or unit question to remind students why they are conducting this lab. Run through any pertinent directions. Explain what students should do if they have a question about something as they work. Define the end-of-investigation outcomes. Usually, this will amount to students completing their lab paper and writing their conclusion. If the investigation will run longer than one class period, explain where students should be in their investigations and by what time. As the lab continues, let students know how much time remains and what they should have accomplished by each time check. Stop work at an appropriate time to allow students to clean up and return equipment to an established location in the classroom. Also, have a plan for collecting or storing student work. Some students are responsible enough to keep their lab sheets, but other students may need the teacher to collect them and redistribute them for review the next class.

# Integrating Writing with Explore

The Common Core State Standards (2012) have specific expectations related to the use of writing in the science classroom. In the primary grades (K–2), students conduct basic shared research projects and write about experiences. Students are also expected to record science observations. In the intermediate grades (3–5), students begin to conduct short research projects independently, take notes, and write routinely for a range of tasks. At the middle and high school levels, students are expected to write argumentative pieces that focus on discipline-specific topics, explain scientific procedures and experiments, and conduct extended and thorough research projects, and to write routinely for a variety of purposes. The Explore phase of the 5E model is the perfect time to provide students with the instruction, support, and tools they need to communicate effectively in science.

Any and all inquiry-based activities should include a lab sheet of some sort that lists each step of the procedure, providing room for students to think, reflect, observe, and respond in writing. During inquiry-based activities, students can record predictions, thoughts, ideas, or questions they have before beginning a particular investigation. During the investigation, one student can be in charge of writing additional questions that arise. Then, the class may discuss these questions, following the investigation. Students collect and record data and then make comparisons and draw conclusions. This step in the process may also be completed in writing on the lab sheet. Finally, model how to summarize and reflect upon what they learned from conducting a particular experiment. This may be completed on a lab sheet, in a science notebook or journal, on index cards, sticky notes, or on any type of paper. Students' written work allows the teacher to ascertain which students met the learning objectives and which students need additional instructional support.

During noninquiry-based activities, students can also write as a natural part of scientific exploration. They may write predictions, thoughts, ideas, or questions related to a topic, skill, or task. When their ideas are shared with the class and posted for other students to see, students know their ideas are valued by the teacher. In the "Digging In" activity, for example, students recorded their observations in a matrix and then used their observations to summarize how the soil samples compared. They could have also made inferences as to why the soil samples were different. During the "Presto-Change-O" activity, students wrote their predictions as to what might happen at each station before the investigations began. They could record the pre- and post-outcomes, then define each outcome after the Explain phase. Once the record sheet was completed, students could again summarize the activity by writing what they learned. During the "Times Change" activity, Mrs. Pike used the sticky note strategy for students to record their ideas before conducting the online simulation. Since most of the work the students were completing was online, they would not have an opportunity to write during the simulation. However, they could record questions they had about natural selection and complete a concept map to explain natural selection.

Students have many opportunities to write in response to scientific learning. The Explore phase of the 5E model lends itself perfectly to this. Students can take notes, record observations, make predictions, summarize their before-and-after ideas, and write conclusions based on evidence. These are all essential skills that teachers continue to build as students work toward becoming scientifically literate.

## Conclusion

The Explore phase of the 5E instructional model allows students to have common experiences in order to make sense of what can sometimes be abstract and confusing science content. This Activity-Before-Content (ABC) approach also supports science application skills, such as predicting, comparing, inferring, and concluding, just to name a few. Allowing students to manipulate and explore science content before actually teaching them about it helps build background knowledge, an essential component of an effective instructional plan.

Conducting hands-on activities involves a lot more planning and preparation than simply saying to students, "Ready, set, go!" First, teachers should decide whether the objectives of the lesson require a true inquiry-based activity, or if students will be conducting a noninquiry-based activity. If the activity is inquiry-based, the teacher then needs to decide if it will be teacher- or student-directed, or a combination of both. And to ensure suitable student-learning outcomes, inquiry-based investigations should follow the steps of the scientific method. Noninquiry-based activities can be just as effective in bringing experiences to students. In either instance, well-planned lessons and well-organized science classrooms help maintain order and allow students to productively attend to the task set before them.

## Questions for Reflection

1. What level of scientific inquiry is most common in your classroom? How might you vary the inquiry levels of the activities conducted by your students?

2. Compare each type of hands-on activity: inquiry-based and noninquiry-based. Which activity will you use more or less of? Why?

3. What is your greatest organizational challenge in conducting science investigations? How might the suggestions in this chapter help your labs run more smoothly?

# Chapter 4

# Phase 3: Explain

The Explain phase of the 5E model allows students to learn the information they need to make sense of their scientific explorations. James Trefil and Wanda O'Brien-Trefil (2009) say it plainly: "If you expect your students to understand molecular biology, you have to teach them about molecular biology" (32). Up to this point in the 5E instructional model, students have been learning inductively. They have questioned, probed, investigated, analyzed, and explored whatever concept or idea the teacher has identified as "next to learn" along their path toward scientific literacy, and they have constructed their own understanding. By now, a teacher might be asking, "So when do I actually get to teach something?" Rest assured. Students have been learning plenty, assuming that the Engage and Explore activities provided students with opportunities to engage with and investigate the world around them. Now that they have common experiences related to the topic, they can learn the facts and transform this information into concrete knowledge. Presenting students with facts and information may take different forms. Each of the following forms is discussed in this chapter:

- reading to learn science

- listening to learn science

- discussing to learn science

# Reading to Learn Science

The following is a note about "Range and Content of Student Reading" from the Common Core State Standards (2012, 10):

> "To build a foundation for college and career readiness, students must read widely and deeply from among a broad range of high-quality, increasingly challenging literary and informational texts….By reading texts in…science, and other disciplines, students build a foundation of knowledge in these fields that will also give them the background to be better readers in all content areas. Students can only gain this foundation when the curriculum is intentionally and coherently structured to develop rich content knowledge within and across grades. Students also acquire the habits of reading independently and closely, which are essential to their future success."

*Informational literacy* is the ability to access, evaluate, organize, and use information from a variety of sources. For science students, this usually means reading informational (nonfiction) texts in order to extract facts and information, which students use (process) to create knowledge. For example, consider this fact: Hurricane Andrew produced an estimated $30 billion in damages. An interesting fact. But what now? Without a purpose to anchor this fact, it remains just that: a fact. But what if this fact arose during a discussion about natural disasters in the United States? Another student might contribute this fact: Roughly 1,300 people lost their lives as a result of Hurricane Katrina, while only 65 people died in Hurricane Andrew. To anchor these facts, students may debate the criteria by which natural disasters are evaluated. To engage students with this topic, a teacher might ask, "Which was a more destructive natural disaster: Hurricane Andrew or Hurricane Katrina?"

Today, teachers have a multitude of options when selecting informational texts for their students, including textbooks, informational articles from credible journals and periodicals, leveled texts and readers, and trade books. Ideally, teachers use a combination of text resources to help students identify relevant information and begin constructing knowledge about scientific concepts from several sources.

## Textbooks

Textbooks have long served as traditional informational resources. Within textbooks, students find information organized by unit, topic, and subtopic, as defined by the chapter titles, headings, and subheadings. They cover a broad range of concepts and topics and can explain them in minute detail. Pictures, graphics, charts, tables, and other visuals may support the text, but the bulk of the information is contained within the text itself. The higher the grade level, the more detailed the text tends to be. Students in third grade might read about how people's actions change ecosystems. Their textbook might include facts regarding construction, pollution, setting fires, planting trees, or irrigation. In sixth-grade textbooks, students might read about specific species (their populations, their communities, and their ecosystems) and the limiting factors (food, water, light, and space) and relationships (competition, predation, and symbiosis) that affect populations. In eighth-grade textbooks, students might read about how energy travels through an ecosystem. They might apply the laws of conservation of mass and energy to ecosystems and read about the carbon cycle. Each grade level's informational texts should build upon the informational texts students read in earlier grade levels.

Although they have been the mainstay of content-area instruction, textbooks are not the end-all source of information for students. Textbooks present mostly facts and details, rarely making connections to the personal lives of students. Textbook reading levels are sometimes above where most students comfortably comprehend nonfiction text, and the informational vocabulary may seem foreign to many students. Fortunately, there are other informational options available for students to read about necessary content.

## Articles

Articles tend to be good sources of information, but they also tend to use more real-life examples. They may begin with a story or a real-life connection, then embed facts and information within the construct of the narrative. For example, students might read about an ordinary person (who may also be a geologist) who studies the position of rock formations to determine if it has been changed by an earthquake. The article about the geologist and his or her work provides students with information about the changing earth and how understanding these processes helps scientists learn more about Earth's history. Students may find this type of reading more palatable and approachable, rather than the dryness of simply opening their textbooks to page 324 and reading about the processes that change the surface of the earth.

Articles are great resources to bridge gaps in content knowledge when the main text (e.g., textbook) may lack information, doesn't go into enough detail, or doesn't cover a particular aspect of a concept that is required for students to learn. Articles can be found in highly respected, credible, and student-friendly publications that support science learning, such as those listed below. Teachers can also find meaningful and appropriate articles written at a range of readability levels online.

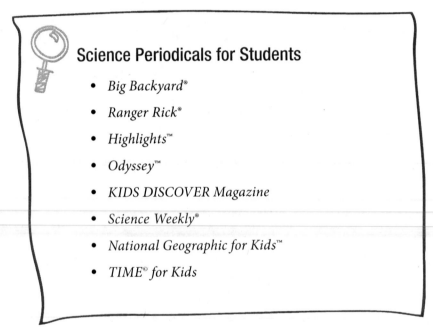

### Science Periodicals for Students

- *Big Backyard*®
- *Ranger Rick*®
- *Highlights*™
- *Odyssey*™
- *KIDS DISCOVER Magazine*
- *Science Weekly*®
- *National Geographic for Kids*™
- *TIME*© *for Kids*

Additionally, newspaper articles provide students with current, relevant, and interesting information related to their community and the world around them. For example, reporters might share the most current information with the public related to local water quality, mysterious animal species, or the aurora borealis. Many newspaper articles may be read online for free. By printing reader-friendly versions of the article, teachers can provide each student with his or her own copy to write on when applying critical reading strategies, such as highlighting, circling, or writing questions in the margins to ask during class discussions.

## Leveled Texts and Readers

Many teachers organize their science reading instruction in a whole-group model. All students, regardless of their reading abilities, have the same text, and the teacher guides the class as all students follow along with their individual texts. But not all texts are equally effective for all students. A teacher may want to have the majority of students reading one text independently or in pairs while a small group of students with similar reading abilities reads a related text more closely matched to their instructional reading levels. This strategy is often referred to as "Guided Reading" (Fountas and Pinnell 2001). During this small-group time, the teacher guides the reading instruction as students read silently or aloud. The teacher also models and demonstrates appropriate reading strategies, such as identifying main ideas and details, retelling, summarizing, questioning, previewing, or identifying cause-effect relationships. Since the group is small (typically three to eight students), the teacher can more closely monitor students' reading comprehension and conceptual understanding.

To truly utilize a guided reading approach, students should be reading materials that closely match their reading levels. For this purpose, teachers can use leveled texts or leveled readers, sometimes referred to as guided readers. Some textbooks provide guided reading support as part of their core science program. Usually, guided readers will closely match the content from each grade level's textbook. Some guided readers are published independent of textbook companies. Many come with lesson plans and implementation suggestions, activities, and assessment options.

## Trade Books

Let's not forget that authors have been publishing nonfiction trade books about countless science-related topics for innumerable years. School and public libraries alike have shelves filled with such nonfiction selections, books students may find interesting and enjoy reading. These books also come in a variety of reading levels so that students at varying abilities can read and enjoy them.

Some books read like fictional stories but are entirely nonfiction. Many include cartoon-like illustrations that strongly support the text and set the tone to be kid friendly. Books by Gail Gibbons, for example, are engaging and educational. Some nonfiction books read like kid-friendly encyclopedias. They are filled with valuable information on pertinent science-related topics. Students do not necessarily need to read these books from cover to cover. They have eye-catching photos and illustrations that capitalize on students' visual interests, leading them to voluntarily read the informational text that accompanies them. The Dorling Kindersley (DK) *Eyewonder* and *Eyewitness* series fall into this category. Many fiction authors publish trade books with science-related themes that support and enhance the informational learning that takes place in the classroom. These books are what educators refer to as picture books. They have more pictures than text, and they typically are one simple story. There are fiction chapter books that work in a similar way as well. *Bread and Jam for Frances* by Russell Hoban (1969) and *Island of the Blue Dolphins* by Scott O'Dell (1984) are good examples of fiction chapter books. When using well-developed Engage and Explore activities effectively, supplemented with factual text, teachers can develop highly engaging lessons to support science content. Sometimes trade books fall somewhere in between fiction and nonfiction. These books present facts and information to students in a kid-friendly, inviting, and nonthreatening manner. They can be valuable, engaging informational resources. *The Very Hungry Caterpillar* by Eric Carle (1994) and the *Magic School Bus* series by Joanna Cole are examples of this.

Regardless of the style, trade books are powerful tools for the Explain phase and support students' learning and understanding of scientific concepts.

**Recommended Books!**
Search the National Science Teachers Association (NSTA) website's "Outstanding Science Trade Books" section for a list of great books.

## Listening to Learn Science

Text resources are a valuable and integral part of any content-rich curriculum, but other modes of learning can supplement the reading material. There are other means of bringing content to students. Audio-visual presentations and lectures are also effective informational delivery techniques.

## Audio-Visual Presentations

Audio-visual presentations can take many forms, such as posters, charts, audio recordings, films, and video clips. Students can gain insight and information related to a number of concepts by watching relevant and appropriate audio-visual presentations on scientific content. For example, short videos were suggested as an effective resource for the Engage phase. Now, during the Explain phase, students can watch longer video segments that provide a more comprehensive overview of the content. Specifically, as an Engage activity, students might first watch a short video clip related to plant and animal adaptations or observe various images of plants and animals to identify their characteristics that enable them to survive in their environment. For the Explain phase, students can watch a longer, more comprehensive video related to the adaptations of desert plants and animals that allow them to survive in a harsh climate. To internalize their learning, students can record the adaptations beside the name of each plant or animal mentioned in the video, following the outline of a Two-Column Chart notes page (see Appendix C). Audio-visual resources are available from any number of reputable production companies. Or if a school has limited resources or funding, online resources are plentiful, and many do not require logins or fees. Many resources can be downloaded and saved as computer files for easy viewing at a later date.

## Lectures

During the Explain phase, teachers can share their expertise, explain important or detailed concepts, and clarify information through lectures. Today's lectures take many forms, including monologue, conversation, digital delivery, or interactive presentations, to name a few. Sometimes, lectures include elements of all of these sharing strategies. Teachers can find premade slideshow presentations if they prefer this delivery method, or they can put together their own dynamite presentations using any number of presentation software programs. These presentations, once created, are ready for classroom use, and then, from year to year, they are easily edited and revised. They are also easily shared among science or grade level teams, so dividing and conquering this type of work facilitates lesson planning for everyone.

Oftentimes, students seem to listen better when someone else shares information. Having guest speakers is another means of bringing content to students through lecture in a novel fashion. Electricians, parks and recreation service workers, zookeepers, ecologists, and water treatment plant operators can all bring the outside world to life inside the classroom. Some guest speakers may not be available in person, or some may live too far from the school. Teachers can use digital resources to bring experts right to their classrooms for "guest lectures." Podcasts, digital audio files, and online video presentations are effective means of bringing experts to students without them having to be physically present. One such online resource is *The Scientific American*™ (http://www.scientificamerican.com). *The Scientific American*™ publishes podcasts regularly. Some are several minutes long, but those listed under "60-second Science" last just one minute. Likewise, YouTube™ (http://www.youtube.com) and TeacherTube™ (http://www.teachertube.com) are websites that provide ready access to informational videos on virtually any topic. All teachers need to do to find the right presentation is browse the videos and select one appropriate for what students are studying. Another option for integrating technology into lecture is video chatting. Video chats have a live video feed from computer to computer. People can talk to each other from down the street or across the globe when they are connected through video chat. Rural schools could benefit from a meteorologist's presentation through video chat. Or perhaps there is a science research laboratory that is too far away for the lab technicians to visit the class. The class could connect with the lab technicians by video chatting, listening to their presentation, and then asking questions.

## Discussing to Learn Science

After students have read about and listened to information about scientific concepts, leading discussions can reinforce the learning that has taken place so far, which allows students to clarify questions and uncertainties they may still have. This may be accomplished in any number of ways. Teachers may engage students in whole-class discussions. For the most part, the teacher asks questions and students volunteer responses. Although common, whole-class discussion is only engaging for the students who participate. Other techniques, such as paired or small-group discussions, allow more students to talk in the same amount of time. This way, students hear more than one student's perspective, and they have opportunities to assimilate others' ideas into their own (or debate them, which is an even more complex task and really gets students personally involved with science). To ensure that everyone has equitable talk time, students can each take on a role in the group discussion, such as those described on the following page.

Depending on the group size, the teacher may choose to use all or some of these roles or think of additional roles. By assigning roles, students hold each other accountable for their ideas, and everyone has the benefit of participating. Another way to ensure student participation is having students choose a number between one and six and take turns rolling a die. The student whose number is rolled must answer the question.

## Small-Group Discussion Roles

- **Clarifier:** Restates the teacher's question using different words

- **Questioner:** Asks, "Why do you think this?" after each student responds to the teacher's questions

- **Encourager:** Gives positive feedback to the person responding *("I like your idea about...")*

- **Connecter:** Makes connections between two or more people's ideas

- **Summarizer:** Restates the main points of the group's ideas

- **Reporter:** Shares the group's ideas with the whole class

# Vocabulary Instruction

Science vocabulary is very specific. Direct vocabulary instruction increases understanding and consequently increases achievement. Robert Marzano (2004) states that the impact of direct vocabulary instruction can be as much as a 33 percentile point gain on standardized tests. By having direct vocabulary instruction as a regular part of learning, teachers should observe students experiencing success on curriculum-based measures within the classroom.

Students will rarely use *lithosphere, joules,* or *Golgi body* as part of their everyday conversations with their peers. Other words, such as *pitch* and *light,* have numerous meanings in conversational English, but students will likely not use the specific meanings when describing scientific processes or occurrences. These specific and multiple-meaning terms can be especially confusing to students with language disabilities or to English language learners. Direct vocabulary instruction is an essential component of any science program. It fits well in the Explain phase of the 5E model. The following suggestions can be used to help students learn scientific vocabulary:

- definitions and examples

- nonlinguistic representations

- word parts

- word practice

## Definitions and Examples

One method of teaching students new terms is by explicitly providing them with the definitions or by having students look up definitions in a glossary or dictionary. Students can record these definitions in a notebook or a personal science dictionary. One factor to consider about this method of vocabulary instruction is that dictionary definitions can be highly academic and use language that is abstract and difficult to grasp. If students lack adequate background knowledge or the definition is above their reading level, little learning can occur. Beck, McKeown, and Kucan (2002) suggest having students write explanations of definitions in everyday language. See Figure 4.1 for an example.

**Figure 4.1** Student Vocabulary Definition and Example

| Term | Dictionary Definition | Explanation |
|------|----------------------|-------------|
| lithosphere | the rocky and solid outermost layer of the earth; includes the crust and the upper part of the mantle | This is the top layer of the earth. It is rocky and solid, not liquid like the mantle. It has the crust and the hard upper mantle. |

## Nonlinguistic Representations

They say a picture is worth a thousand words. Teachers can put this saying into practice as an effective vocabulary learning strategy. Marzano (2004) suggests having students represent words nonlinguistically (e.g., through pictures, graphics, and physical motions). Students who are learning about plate tectonics might illustrate each type of plate boundary beside its definition or explanation. Or students might develop a hand motion to demonstrate each boundary, such as sliding their hands past each other (transform boundary), pushing their hands together so that one slides over the other (convergent boundary), and pulling their hands apart (divergent boundary). Whole-body motion can even be used. An example of this is having students be trees in order to demonstrate the vocabulary associated with plant parts. They start off rolled up in a ball (seed), then one foot sticks out (the root), then an arm pops up (the stem and leaves), then the whole body moves up (trunk and branches).

## Word Parts

Direct vocabulary instruction provides an opportunity for teachers to help students become both independent thinkers and readers. Teaching students about word parts like Greek and Latin prefixes, suffixes, and roots will allow them to read, analyze, and understand unknown words in context. Science is especially filled with words that can be broken down into smaller word parts, analyzed, and understood. For example, students who know that *litho-* means "stone" or "rock," and that *sphere* means "ball-shaped" will better understand words with similar prefixes and roots when they encounter them in unfamiliar texts. And knowing the meanings behind these word parts will help students better remember their definitions, thus leading to increased scientific literacy (Rasinski et al. 2008).

## Word Practice

Once teachers have directly instructed students on the essential scientific vocabulary, students stand a better chance of remembering the new vocabulary the more they use the words. Anyone who has tried to commit to a word-of-the-day or word-of-the-week program can attest to the fact that unless the word is used regularly, it is rarely committed to memory. To help new vocabulary become a working part of students' understandings, provide students with opportunities to use and practice the words through discussion and word play games.

Discussion helps students internalize information. Once students have learned the definitions and provided their own explanations, they may connect and apply these words to the Engage and Explore activities. The teacher can lead a class discussion by asking, "How are Term 1, Term 2, and Term 3 related to the activity when we _____?" Or for paired discussion, teachers can have students turn to partners to review the words they learned, use them in sentences, and reflect on how they apply to previous activities. In small groups, students can rate each word with a 5, 3, or 1 (5 being "I know it well," 3 being "I kind of understand it," and 1 being "I don't know what this word means"). After they rate the words, they can collaboratively review and discuss the ones that are the most challenging, seek clarification, or use their text resources to learn more about the terms and how they apply to the concepts they are studying.

Playing games is a useful strategy to help students meaningfully practice using terms in a nonthreatening environment. To play "Vocabulary Game Pyramid," teachers create a six-word pyramid following the model in Appendix C. Then, students pair up. One student faces the words, either on paper, projected on a screen, or on the board, and gives the other student clues about the word he or she is trying to guess. The second student cannot see the words. The clue-giver gives clues to the responder so that he or she can correctly identify the word. The objective is to correctly identify all six words in a specific time frame. Another vocabulary game gets students up and moving. The teacher tapes a vocabulary word to each student's back without them seeing what the words are. Then, students move about the room, asking *Yes* or *No* questions about their word in an attempt to correctly identify the word. Once they know their word, they sit down.

## Supporting Science Reading

Nonfiction reading can be a difficult task for some students. Nonfiction reading requires different skills than reading fiction. Students read nonfiction for a specific educational purpose. As Stephanie Harvey (1998) reminds us, the "structure, content, and purpose" of nonfiction text is different from the fiction varieties that students are used to (71). It follows a predictable pattern, but one that students may not recognize from their fiction reading. And having to interpret and apply scientific data, information, or science-specific vocabulary can interfere with a student's ability to independently read and understand nonfiction texts. For these reasons, apply simple reading strategies that support and scaffold learning as students attempt to make sense of what can sometimes be confusing text.

## Set a Clear Purpose for Learning

Students generally read nonfiction texts to learn information, which is a different purpose from reading fictional texts. As with any reading selection, students should know the reason they are reading a particular text. In science, these reasons could include learning facts, learning about a particular process, finding out detailed information about a scientific concept, or making a connection between the world of science and the world in general. Think of the purpose of the reading as an itinerary, and the text as a road map. It is important to know where the learning is headed before pulling out the map to look at it. Setting a purpose for reading provides students with a learning destination. Then, when they begin reading the text, they know where the map is going to take them.

## Teach Text Structures

Stephanie Harvey (1998) supports the need for readers to understand expository text structures since they vary so greatly from narrative frameworks and the elements of story. Expository text structures include cause-effect, problem-solution, question-answer, compare-contrast, description, sequence, and main idea. Identifying and analyzing nonfiction text structures can help students better comprehend the text. As Harvey notes, "If students know what to look for in terms of text structure, they grasp the meaning more easily" (7). Of course, the ultimate goal of teaching students about text structures is for them to independently apply this learning when they read informational text during future learning opportunities in order to support comprehension.

## Teach Text Features

One challenge students face when encountering the language of informational texts is that they are unfamiliar with the layout and features of the text. Nonfiction authors tend to use unique characteristics to direct and focus the readers' attention to important information. Teachers can help students overcome these challenges by teaching, modeling, or illustrating how the text features can help students predict and clarify information as they read. Students should know how text features, such as headings and subheadings, bold or italicized text, and captions, organize, emphasize, or clarify information. This can help students when they read independently.

## Differentiate Reading Instruction

Another challenge students face when they read nonfiction is that the language itself may be at a readability level well above students' own instructional or independent abilities. A nonfiction excerpt from a seventh-grade textbook, for example, might be written at a ninth-grade reading level. Determining the reading level is a complex process that utilizes formulas that take into consideration vocabulary, sentence length, and sentence structure. Any one or all of these factors might challenge students' abilities to read and comprehend the subject matter. Teachers can help students overcome this challenge by utilizing differentiated reading strategies in their science classrooms and selecting informational texts appropriate for students' reading abilities. These strategies include:

- **Chapter Walk:** Before beginning a new chapter, preview the chapter as a class. Discuss the main text features throughout the chapter and have students make predictions based on those text features about what will be covered and what the main ideas may be.

- **Stop and Discuss:** After every few paragraphs or sections, stop reading in order to have students retell facts, paraphrase, or summarize the content.

- **Leveled/Alternate Texts:** Provide students who read below grade level with an altered or leveled version of the text, one which includes the essential information but is written at a readability level better suited for students' abilities.

With direct instruction and appropriate support, students can be successful with challenging content they may encounter in their nonfiction reading.


# Integrating Writing with Explain

The Explain phase of the 5E model offers opportunities for students to write about their learning. Students can better secure the information they learn when they record ideas and summaries in a notebook, on paper, or in an electronic document. This allows students to tangibly retrieve the information they learned when they need it.

One way for students to write during the Explain phase is by taking notes. Students' notes can be stored in a notebook, a science journal, or a pocket folder. Paper may be lined, blank, or show a graphic organizer. Whatever the format, students who take notes have a permanent record of their informational learning. The following are suggestions for student note-taking during the Explain phase:

- record information and facts learned from lecture or reading

- illustrate processes using diagrams and labels

- list, sort, categorize, and identify unique characteristics to compare concepts and ideas

- summarize results of investigations and connect them to informational text

- write, explain, and illustrate vocabulary terms

Figure 4.2 shows a fifth-grade student's notebook page with notes recorded during a study of the water cycle.

## Figure 4.2 Student Notebook Entry

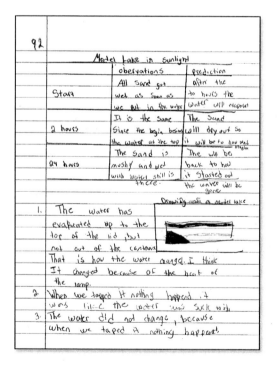

Another way students can write during the Explain phase is through personal reflections. This type of writing can be very powerful for students as they express their thoughts and solidify their understanding. If needed, teachers can provide students with thinking stems to help them begin their writing (McGregor 2007). Sample thinking stems are listed below. Written reflections allow students to think about and reflect upon what they have read and learned and to begin transitioning facts, information, and ideas into knowledge. Teachers may choose to have students write reflections at various points throughout the Explain phase to document the progression of learning they undergo.

## Thinking Stems

- I think…
- This makes me think about…
- I notice…
- I can relate to _____ by _____.
- I wonder what would happen if…
- What is important is…
- I see…
- It's interesting that…
- I feel…
- I used to think _____, but now I think _____.
- I remember…
- I understand…

# Conclusion

The Explain phase of the 5E instructional model is when students learn essential content-area vocabulary, facts, and information. This step along the instructional learning path is usually more teacher-directed than the Engage or Explore phases. Information can come to students from more than just traditional sources like textbooks. Trade books, articles, and informational websites provide students with a wealth of information, usually presented in a more kid-friendly and interesting manner. Teachers can use audio-visual presentations, lectures, and class discussions to help students comprehend new information provided in scientific texts. Podcasts, video chats, and other electronic resources can also be used to reach out to any number of scientific experts to satisfy students' curiosities about any number of concepts.

Teachers should remember that no single strategy meets the needs of all students. Some students will need instructional support to gain access to content, to remember facts, and to convert information to deep understanding. This may be accomplished through direct instruction of text structures and text features. Or teachers may need to provide text with lower (or higher) readability levels to support students on their paths to becoming scientifically literate.

## Questions for Reflection

1. What type of nonfiction reading is most prevalent in your classroom? What additional types of nonfiction reading might you use to develop students' understanding of scientific concepts?

2. What text structures are present in the literature that students are reading? How will you teach about text structures to help students navigate nonfiction texts? What text support (notes, graphic organizers, collaboration) might students use to better understand what they are reading?

3. What are some obstacles to providing students with varied levels of science texts? Explain how you might overcome these obstacles in your own science classroom.

# Chapter 5

# Phase 4: Elaborate or Extend

Up to this point of the learning process, students have followed the lead of the teacher. They have engaged with a topic, explored ideas related to the science surrounding the topic, and had the content explained through various delivery methods (e.g., textbooks, lecture, video). By now, students should have a solid understanding of the concepts presented to them. The next phase in the 5E instructional model is Elaborate or Extend. This is where teachers bring related concepts to the forefront and allow students to investigate and explore ideas that interest them or otherwise stretch their learning to new levels.

In this phase of the 5E model, students should already possess foundational knowledge from which elaboration may occur. This phase is when students make connections to other concepts that have naturally arisen during class discussions. The Elaborate phase allows students to more clearly define concepts and ideas in real-life scenarios. Here, students can make learning personal. This can happen when the teacher provides extension activities that are relevant and important to students, or they can be student-guided extensions with teacher support. The Elaborate phase also allows the teacher to more carefully individualize learning, providing independent study projects for students who are above-level learners, teacher-directed activities for on-grade-level learners, remedial work for students who are below-level learners, and additional language support for English language learners.

The vastness of this more open-ended phase of the 5E instructional model may seem overwhelming. But teachers who keep their learning objectives in mind when planning an Elaborate lesson are sure to maintain a focused learning environment with clear expectations. The extensions that follow fit into the following categories:

- whole-class extensions
- small-group extensions
- independent extensions

# Whole-Class Extensions

Teachers may choose to extend learning through a whole-class delivery method. This is useful for conducting a teacher-directed lesson or when the concept is solidly understood. These activities are best suited for whole-class extensions:

- interactive investigations
- application extensions
- hands-on simulations
- debates

## Interactive Investigations

The use of technology is a motivating and engaging strategy. Lemke and Coughlin (2009) encourage educators to "leverage the opportunities offered by Web 2.0 technologies" (58). Using technology as a means to explore additional information about a scientific concept is an effective way to promote self-directed learning and to get students interested in the process of digging deeper. In interactive investigations, students and teachers search for videos, online articles, images, and other websites that elaborate and extend content from the Explain phase. For example, students studying the relationship between parasites and hosts may extend their learning by investigating the harmful effects of parasites on humans. To do this as a whole class, conduct a search for maps, articles, and data about how mosquitoes transmit malaria around the globe. The amount of information found and the way it is searched for can be determined by the age of the students and the time constraints in the classroom. The sky is the limit when technology is at the core of extending student learning.

## Application Extensions

As students approach the end of a unit, they may participate in an extension activity that asks them to apply their learning to a new concept. This type of extension activity would directly support the learning that has already taken place. For example, students who have learned about the life cycle of the butterfly may now, during the Elaborate phase, learn about the life cycle of a frog. Students may read a book about the life cycle of the frog as a whole class, in small groups, or independently, and then identify similarities and differences between the two animals' life cycles. This comparison serves to strengthen and solidify students' understanding of the life cycle of the butterfly. In another example, students learning about closed circuits might learn about parallel and series circuits. In order to demonstrate parallel or series circuits, students will inevitably demonstrate their understanding of closed circuits. Although these topics are similar, the objectives differ, and students must extend their general understanding of closed circuits in a different context.

## Hands-On Simulations

Simulations are fictitious activities that represent real-life events. In the Elaborate phase, simulations extend the learning that has taken place around a particular scientific concept. These simulations can range from simple to complex, depending on time and student readiness. Ideas for simulations in each field of science are provided in Figure 5.1.

**Figure 5.1** Simulation Ideas for the Elaborate Phase

| Science Area | Grades K–2 |
|---|---|
| **Life Science** | Students can act out the life cycle stages of a butterfly. |
| **Earth Science** | Students can act out day and night. When the teacher turns on the classroom lights (to replicate daytime), students get up and move about safely, engaging in activities they perform throughout the day. When the teacher turns off the classroom lights (to replicate nighttime), students rest at their desks, pretending to sleep. The teacher can use both settings to discuss what students see in the sky at each particular time of day. |
| **Physical Science** | Students can make a paper cup "roar" by decorating the open end to look like an animal that roars. Then, insert a piece of string through a small hole in the bottom of the cup. The end of the string inside the cup should be anchored with a paperclip. When students pull the string between their fingers, the cup seems to make a roaring sound. |

| Science Area | Grades 3–5 |
|---|---|
| **Life Science** | Students can use cotton swabs (to act as a bee's legs) and colored baby powder (to act as pollen) placed in students' hands (to act as the flowers) to simulate how bees pollinate flowers. A student holding two cotton swabs "flies" from "flower to flower," picking up pollen from one plant and carrying it to another. |
| **Earth Science** | Students can make salt, alum, or sugar crystals to simulate the geologic process of crystal growth. |
| **Physical Science** | Students can use paper and tape to build a model neighborhood on a school sidewalk or secured area of the parking lot. Then, the teacher can use an electric or gas blower to simulate hurricane force winds or a tornado. Students can record the pre- and post-wind effects in science journals. Back in the classroom, they can compare their observations and draw conclusions about the effects of high winds. |

| Science Area | Grades 6–12 |
|---|---|
| Life Science | The teacher can place colored toothpicks, each representing a different woodland animal, in and around a certain area of foliage on the school campus. Students can work independently to count the number of each color toothpick they see and draw a map of where they find them. Back in the classroom, students can collaborate and compare the quantities of each color of toothpick they observed and learn the total of each "species" from the teacher. This can then lead to a class discussion about animal adaptations and what kind of animal each toothpick might represent in the simulation area. |
| Earth Science | Students can set up models of the earth using plastic bins, sand, soil, rocks, and other natural materials, then set them outside in the open elements. Students can take pictures of the earth for five to ten days and put a slideshow together to visually demonstrate the natural changes that occur to the earth over time. |
| Physical Science | Students can use data collected from the Chandra X-ray Observatory to calculate the average pixel intensity of X-ray emissions from a supernova remnant. (See NASA's website for more information.) |

## Debates

Nothing gets the blood pumping and students personally involved like a good debate! Debating definitely heightens positive emotional commitment, and debates are highly interesting for both the debaters and the audience (Marzano and Pickering 2011). Everyone can participate in a debate in some way, either as the speaker, as a researcher, or as an inquisitive audience member. Amy Azzam (2008) supports the use of debates since they serve as platforms for learning and practicing critical thinking skills, develop students' organizational skills, provide meaning and purpose for literacy through research, and empower students "by giving them a voice; [debates] can transform them into advocates for themselves and their communities" (69).

During the Elaborate phase, students can research the topic and formalize their thinking in order to support their positions with facts, information, statistics, and other evidence.

## Steps to an Effective Debate

1. Choose a topic.

2. Take (or assign) a side.

3. Research talking points to support and refute each side.

4. Select a moderator. This person will lead the debate. He or she will give directions, introduce the topic and speakers, and inform each speaker when it is his or her turn to talk. (Another option is having each student research at least one pro and one con about the topic, followed by a structured class debate about the topic.)

5. Have the audience listen attentively and take notes on each side of the debate. Allow time for an audience question-and-answer session. Students can direct their questions to either one or both of the speakers.

Depending on students' interest in the topic, the importance of the topic, and the teacher's ability to devote adequate time to the task, a debate can take on a formal or informal construct. As a formal event, students dress appropriately. There is an official moderator. The audience respectfully follows expectations and is silent when others speak. Audience members also follow a formal process for asking questions. Or teachers can opt to conduct more informal debates with students dressing and acting more like themselves. The debate itself should follow certain expectations, including mutual respect from both speakers and audience members. However, the structure of the debate process may be more casual. The teacher may decide to have all students take the podium to make their positions known. Or students may work as teams to organize and present their position. Regardless of the structure of the debate, it is considered a whole-class activity since every student in the class plays a role in the process, whether they are a speaker, moderator, or audience member. The teacher should set clear expectations for each student's role and consider how all students will be held accountable for their own personal efforts. For example, if not all students conduct research to support a position and instead act solely as audience members, they may be required to take notes during the debate, perhaps in a two-column chart, to summarize each speaker's points.

# Small-Group Extensions

Small-group extensions allow all students to participate in learning at the same time. These types of activities require students to work collaboratively to complete a group project or task and can be great tools for extending student learning. Some activities for small-group extensions are:

- problem-based learning
- lab explorations

## Problem-Based Learning

Problem-based learning activities require students to design, create, and resolve a current and real problem. Students work in small groups to assess a problem and apply what they've learned from the previous phases to solve the problem. For example, students may have learned about erosion by creating landforms with sand and soil and observing the effects of wind and water. A problem-based learning activity related to the landform investigation might read as follows:

*Your community is concerned about erosion near lakes and rivers. Design and construct a means of minimizing the effects of erosion by wind or water.*

To complete the task, students would research methods of lessening the effects of erosion and design a prototype to test for this purpose. If feasible, students would construct their prototype and test it, gather data and analyze it, and then reflect on their idea. They would suggest improvements for future tests. This activity, although similar to a lab investigation, requires students to use science, technology, engineering, and mathematics concepts and skills to complete from start to finish. This type of activity challenges students' thinking and requires them to apply their knowledge for a useful purpose.

When engaged in a problem-based learning activity, students consider a problem, then plan and execute a proposal to solve it. Students use interdisciplinary knowledge and skills to do this and therefore extend their understanding of the original concept being studied.

## Lab Explorations

Teachers who follow the 5E model have previously provided their students with meaningful and engaging hands-on activities during the Explore phase. However, sometimes just one hands-on activity is not enough. The purpose of conducting a whole-class, hands-on activity during the Explore phase is to allow students to have common experiences with the concepts they will learn about. It is a means of building background information and introducing key vocabulary and overarching concepts through hands-on, meaningful experiences. Using additional hands-on activities during the Elaborate phase will help students strengthen and internalize the ideas they have been learning about. It will give them an opportunity to use what they have learned and apply it beyond the limits of the learning objectives. For example, intermediate students who conducted an investigation related to sound energy during the Explore phase might now conduct an investigation related to seismic activity. This Elaborate activity may be similar to the one conducted earlier, but now students are applying their knowledge of sound energy to a real-world topic.

An Elaborate activity might also be an extension of an investigation that was set up and completed during the Explore phase. For example, primary students who investigated whether worms are sensitive to light or to touch might now record observations of the worms in their natural habitat (worm farm) for several days. During the Elaborate phase, students could summarize their observations about worms in their habitats as an extension of their investigation during the Explore phase.

# Independent Extensions

An effective means of engaging all students in a project or assignment is requiring specific work from each student. The 5E instructional model allows room for independent projects and activities during the Elaborate phase. Some ideas for independent extensions are:

- independent projects
- physical representations and models

## Independent Projects

The extent to which students commit to a topic is, in part, a function of their level of interest (Marzano and Pickering 2011). Teachers can elevate students' interest level in a topic by assigning a research project that allows students to decide what they will study. Oftentimes, teachers focus on one particular objective within a scientific concept. But as questions arise during class discussion, many teachers tuck those away to discuss at a future date. The trouble is, with so much content to cover, that future date may never arrive. Assigning related topics as independent projects is a useful means of allowing students to extend their own learning by studying related topics of interest to them. For example, a teacher whose fourth-grade class learns about sound energy may not have time to learn about related topics, such as sonar, echolocation, whale communication, the Doppler Effect, sonic booms, sound proofing, the mechanics behind synthesizers or audio speakers, white noise, the history of the gramophone, etc. However, the teacher can assign these topics to students according to their interests during the Elaborate phase. Students can research information about the topic and summarize it by creating a slideshow presentation, a simple written report, a poster, or other viable means. Then, each student can share the information he or she discovered about each topic with the class. This way, the teacher has helped students master the objectives of the lesson, and students have extended their learning by researching and sharing information about related topics of interest.

## Helpful Tips for Student Success with Independent Study Projects

- **Choose a topic of interest.** Have an adequate pool of topics from which students may choose their independent study project. Decide whether more than one student can choose the same topic.

- **Provide appropriate resources.** Students may need assistance finding resources to conduct their research. Review the options students have, including, but not limited to, textbooks, encyclopedias, trade books, credible online resources, or experts they can interview. Additionally, struggling students and English language learners may need text resources at varied reading abilities.

- **Set clear expectations.** Be sure students know how their independent study project will be evaluated before they begin working. Rubrics are a great way to help with this. Also, be sure students know the timeline and due date for the project.

- **Provide instructional support.** Some students may need support researching and identifying the main ideas and details about a topic. Graphic organizers may help to support these students. If needed, students can draw and label an illustration of something related to the topic as an alternative to writing an expository summary.

- **Set goals and check in.** For longer projects, have students set goals along their path of research (e.g., a date by which they will have the research completed or a date by which they will have a draft summary completed). Then, periodically set aside five to ten minutes of class time to check whether students are on track to complete their projects.

## Physical Representations and Models

An effective strategy to help students internalize information, leading to true understanding, is having them visually represent an idea through pictures or illustrations. This idea was first introduced during the Explain phase of 5E as "nonlinguistic representations" (Marzano 2004). A related strategy, but now during the Elaborate phase of 5E, students build physical representations or three-dimensional models to illustrate a concept. This strategy is particularly helpful for struggling students and English language learners since they can construct a concrete representation for the knowledge they have learned. They do not need to depend solely on words to "learn" a particular concept.

Physical models make the abstract concrete. For example, when learning about animal habitats, a teacher may read Eric Carle's *A House for Hermit Crab* (1987) during the Explain phase. Following this reading, students may identify, discuss, and record the animal dwellings they observed within the book. As a learning extension, students can create a model of a sea anemone by using craft supplies or other classroom or household objects. The key to this strategy is allowing students to conceptualize and apply their understanding through physical representations, providing them with the time to discuss and summarize how the physical representation defines the concept.

## Integrating Writing with Elaborate

There are many opportunities to integrate writing into the Elaborate phase. Writing is an inherent aspect of many of the Elaborate activities provided throughout this chapter. Consider the following writing ideas:

- Students can write a position paper (state an opinion) with substantiated support (facts and information) to use during a debate.

- Students can write a report, brochure, slideshow, or poster to summarize research topics.

- Students can write a how-to manual to explain the functions of a device created during a problem-based learning activity.

- Students can write lab investigations (predictions, data analysis, and conclusions).

Additionally, students can use this lesson extension time to reflect on the learning that has happened so far. One viable means of encouraging written reflection is by having students respond to their learning in a science journal. Science journals can house vocabulary terms, predictions, conclusions, comparisons, summaries, and even personal thoughts. Thinking stems are the perfect reflection starters as students begin to make connections, wonder about related ideas, and internalize learning.

# Conclusion

Elaborate extension activities provide teachers with the time they may need to help students build on previous learning, to deepen learning, or to extend student thinking to real-world applications. Elaborate activities may resemble any activity a teacher might use during the previous three phases of the 5E instructional model. However, the purpose of any of these activities is to give students opportunities to apply, analyze, and reflect upon learning, and to create models relative to the topic they are studying. Lesson extensions may be conducted with the whole class or in small groups. Teachers can also use this time to allow students to independently find out more about related topics of interest.

## Questions for Reflection

1. How can Elaborate activities deepen students' scientific literacy? Describe a time when you have experienced this yourself or when you have witnessed this experience in students.

2. Which type of Elaborate activity would best help to meet your students' present needs: whole-class, small-group, or independent? Describe the activity you would like to use in the future. Establish clear learning goals for using this activity. What will students know and be able to do as a result of this Elaborate activity?

# Phase 5: Evaluate

The final phase of the 5E model is the Evaluate phase. In this phase, the teacher determines whether students have met the intended learning outcomes, which includes assessing students' performance and/or understanding of ideas, skills or processes, and their applications. In this phase, teachers also document student mastery or progress toward mastery of the scientific concepts students were to have learned in class. Students may also be part of the assessment process in this phase through a self-evaluation of some sort.

Traditionally, content mastery is determined through teacher-created or published tests. Of course, formal testing has a place in the science classroom, even in the 21st century. Tests have many advantages for both students and teachers. They are easy to give and objectively score. They provide an overall summary of student learning, encompassing a variety of skills, knowledge, and information. When using electronic versions of classroom assessments, an item analysis instantly shows teachers which specific skills students continue to lack, or ones in which the whole class excels.

As valuable and useful as traditional tests are for summative assessment, they are not the only means by which teachers can evaluate student learning. Alternative assessment methods are useful and meaningful, and they often provide a more holistic assessment of student learning through authentic tasks. They can be used as the sole indicator of student achievement during the Evaluate phase, or they can be used in conjunction with a traditional assessment to paint a larger picture of student learning. This chapter will focus on the following alternative assessment strategies:

- performance-based assessments
- problem solving
- project-based learning
- student self-assessment

# Performance-Based Assessments

According to Sandra Schurr (2012), a performance assessment is "an assessment of how well individuals can do something, as opposed to what they know about doing something" (8). Performance-based assessments require the teacher to observe actions performed by each student relevant to the learning objectives. Not all learning outcomes can, nor should be, assessed in this manner, but it is important to consider this assessment strategy where appropriate because it provides students with a direct opportunity to concretely show what they know.

Much of science knowledge requires students to correctly use equipment and correctly identify the process of scientific learning. To effectively assess student understanding of these ideas, the best option may be to have students physically manipulate equipment or observe and identify the various components of an investigation in progress. For example, if teachers want to know if students can correctly use a triple balance scale or microscope, they may want to set up a station in the classroom for students to use these pieces of equipment. The assessment would amount to the teacher observing and evaluating student use of the equipment in a given context. In a similar situation, teachers may wish to see if students can set up an investigation with particular parameters and correctly identify a control or the independent and dependent variables of an investigation in progress. In this case, teachers can set up a mock investigation and ask students to identify each component of the investigation. For example, in a study of weather-related topics, a teacher may gather up the supplies used to test how the amount of snowfall compares to the amount of rainfall. The teacher would call each student over to a table to share the objective of the experiment. Then, students would orally explain how they would set up the experiment and make their predictions about the outcome. They would additionally identify the control and the independent and dependent variables in this experiment. For each student, the teacher would use a checklist to determine mastery of the science objectives being assessed. Figure 6.1 provides a sample checklist.

**Figure 6.1** Sample Performance-Based Assessment Checklist

# Observation Checklist:
# Science Process Skills

Student: _____

❑ Defines a problem

❑ Plans and executes a scientific investigation

❑ Identifies variables

❑ Identifies the control

❑ Collects and organizes data

❑ Interprets data in charts, tables, and graphs

❑ Analyzes information

❑ Makes predictions

❑ Defends conclusions

Additional Notes: _____

_____

# Problem Solving

Every day, adults and students alike are faced with decisions they have to make. Do I wear stripes or solids? Do I get an iced coffee, water, or soda? Do I shoot some hoops or ride my bike? But some decisions are related to science concepts they have studied. These include, but are not limited to, knowing how to use simple machines to accomplish a task, deciding whether to kill the snake that has infiltrated the garage, and being able to plan weekend activities that are dependent on the weather. In a report published by the Organisation for Economic Co-operation and Development (2004), problem-solving skills are defined as "the capacity of students to understand problems situated in novel and cross-curricular settings, to identify relevant information or constraints, to represent possible alternatives or solution paths, to develop solution strategies, and to solve problems and communicate the solutions" (3). They contend that using problem solving to assess students helps educators raise student performance in this critical competency.

In this same report, three types of problem solving are identified. Each type is listed in Figure 6.2, along with a description of the task and an example of each type of task.

**Figure 6.2** Problem-Solving Assessment Types

| Problem-Solving Type | Description | Example | Grade Levels |
|---|---|---|---|
| System and Analysis | Identify the relationships within a system, or design a system to illustrate relationships | Have students think of an outdoor game that uses pushes and pulls. Then, have them draw and label a picture of the game, showing how it uses both pushes and pulls. | 1–2 |
| Decision Making | Choose from a set of options, with restrictions | Have students build a rock display. Students must label each type of rock correctly by reading the rock descriptions. Have students use the Rock Identification Key to correctly identify each rock in the collection. | 4–6 |

| Problem-Solving Type | Description | Example | Grade Levels |
|---|---|---|---|
| Troubleshooting | Identify faults and provide working alternatives | Present students with the following problem: *Lin has several plants that just don't seem to want to grow.* Have students draw and describe possible settings for each of Lin's plants. Vary the locations based on temperature (cool/warm) and light (bright, little light). Have students explain to Lin the process of photosynthesis and why each plant needs a different environment. | 7–8 |

It is important to consider the readability of the problem when using this strategy. When appropriate, problems should be supported with visuals (e.g., tables, charts, or illustrations). Teachers want to know that students can use the information they learned from a unit of study to solve a problem. They do not want to test their reading comprehension skills. If all students are to complete a problem-solving assessment item, all students should be able to understand the problem.

Students should work independently on this type of assessment. Depending on students' ages, it may be appropriate to read the problem aloud and provide them with varying amounts of time to complete the assessment. It is also important to consider the appropriate type of materials provided to solve the problem. If the problem calls for students to draw a diagram, they will need a blank sheet of paper on which to work and perhaps some colored pencils or crayons to illustrate their understanding and ideas.

# Project-Based Learning

Projects are a great way to assess what students understand about a topic. Sandra Schurr (2012) defines project-based assessment as a check of students' "understanding of and proficiency in a topic or subject by allowing them to generate or create a product to show what they know and what they can do, either alone or in cooperation with others" (7). Products can be physical models, illustrations, written reports or projects, or any other student creation. Student work products are evidence of learning. They are what teachers assign to document this evidence. They are what students create in response to learning.

Projects also fit very well as part of the Elaborate phase of the 5E model. So what was learned during the Elaborate phase can now be assessed during the Evaluate phase. Or a different project can be assigned and assessed during the Evaluate phase—it simply depends on the time available and the objectives needing to be assessed to demonstrate mastery of the intended learning outcomes. When assigning projects to evaluate student learning, teachers should adhere to the suggestions that follow. The project ideas listed in Figure 6.3 give students a creative means of sharing their understanding of science concepts. Teachers will have evidence of students learning, and students will enjoy demonstrating their mastery of scientific content through a student-centered, authentic task.

## Suggestions for Assigning Project-Based Assessments

- Begin with a clear objective. What should students know or be able to do?

- Decide on a project format. Every project should include some written component, whether it is an explanation, summary, description, opinion, or narrative. The format should clearly connect to the lesson objective(s).

- Determine how the learning objectives will be assessed and how the overall project will be evaluated. This may take the shape of a rubric or checklist. Each component should be clearly explained to students. A sample rubric for the grades 3–5 activity in Figure 6.3 is included in Appendix B.

- Set reasonable expectations with regard to completion deadlines. Include students in these decisions. Plan time to check in with students, if needed, or allow time for them to share their progress with partners in class.

**Figure 6.3** Sample Project Ideas for Assessing Student Learning

| Grades K–2 | Write a story as if you were a caterpillar. Start your story by telling who you are and where you live. Explain what is happening to you during each stage of your life. Tell how you feel during each stage. What happens at the end of your story? Where do you go? What do you do? |
|---|---|
| Grades 3–5 | Write a song comparing the inner and outer planets. |
| Grades 6–12 | On a poster, draw one of these settings: playground, cafeteria, soccer field, kitchen, game room, or amusement park. Label examples of *work*, *Newton's First, Second,* and *Third Laws*, and *balanced* and *unbalanced forces*. Using captions, explain how each label demonstrates each idea. |

# Student Self-Assessment

One piece of the assessment puzzle that is sometimes overlooked is the students' personal ideas regarding their own learning in science. Teachers want students to think critically. What better way to apply this skill than to have students be active members of their own evaluation process? This can be accomplished by having students reflect upon and summarize their learning during a particular unit of study. This may take the form of a checklist, rubric, open-ended questionnaire, or any combination of these. For example, a teacher may provide students with a series of *Yes* or *No* questions and include written response items at the bottom of the page. An example of a self-assessment rubric is shown in Figure 6.4.

**Figure 6.4** Sample Student Self-Assessment Rubric

# Student Self-Assessment Rubric

Name: _____   Date: _____
                                                      Unit: _____


How did you do during this unit in science?

1. I completed my labs on time.              Yes        No

2. I completed my classwork on time.         Yes        No

3. I read my assignments.                    Yes        No

4. I participated in group projects.         Yes        No

5. I participated in class discussions.      Yes        No

6. I always did my best work.                Yes        No

7. My projects and quizzes/tests show that I learned _____ of the information during this unit.

    ❑ All   ❑ Some   ❑ Most   ❑ Little/None

8. I believe I earned a grade of:   **A**   **B**   **C**   **D**   **F**

   Here's why:

   _____

   _____

9. Something new I learned about this topic was:

   _____

   _____

A critical component to student self-assessment is individual conferencing between teacher and student. Teachers may be surprised at how students assess their own learning. Some students will be too hard on themselves, rating their overall learning lower than their evidence of learning suggests. Others will try to take advantage of this opportunity and rate their learning higher than their evidence of learning suggests. Therefore, the teacher-student conference is important in finding this balance. This should not require more than one to two minutes per student. It may be necessary to meet with each student in the class. A second option could be to review all students' self-assessments individually, then conference just with those students who seem to be inconsistent with their evidence of learning.

Another critical component to consider is how students' own self-assessments factor into the students' final grade. Grading is a complex process. Regardless, students should clearly understand how they will be evaluated prior to beginning the work and understand how their self-assessment factors into that evaluation.

## Other Assessment Opportunities in the 5E Model

Formative assessments are those that the teacher uses throughout a course of study to gauge student learning. They are administered frequently throughout the instructional process in order to inform future instruction. Meaningful formative assessments help teachers determine who is meeting the learning outcomes and who is not. When conducted effectively, the teacher should use the results to plan additional remedial or extended learning opportunities for students, depending on what insights the results provide the teacher.

As described, the 5E instructional model is a cyclical process. Because of this, teachers are not limited in their instruction by attending to each phase in a linear manner. Formative assessment occurs throughout the learning process. As teachers conduct formative assessments of student learning, they can cycle through the various stages of the 5E model to provide the most appropriate instructional support to the students who need it. For example, if a teacher identifies during the Explain phase that students have not had enough experience using critical vocabulary, he or she may provide additional instruction to support students as they continue to learn and use the terms. This may take the form of supplemental text resources, additional Explore investigations, and/or carefully developed Elaborate activities. Or the teacher may choose to cycle through the entire 5E process again with a parallel topic, one that supports the ideas from the initial topic with similar vocabulary.

To maintain the 5E model as a cyclical instructional model, the following formative assessment options may help teachers keep track of their students' conceptual development:

- teacher observation
- student work

These options, in turn, allow teachers to more carefully construct direct, shared, and independent supplements and extensions to meet students' needs and ensure optimal success for all.

## Teacher Observation

Teachers can take in a lot of information about students and what they are and are not learning simply by observing them in the classroom. This happens throughout all phases of the 5E model, and during whole-class, small-group, and individual work time.

Through observations that occur during whole-class discussions, teachers can identify students who tend to stay quiet and those who respond frequently. Through purposeful questioning, teachers can determine if the nonparticipants are bored, confused, or simply lost. They can also identify the level of knowledge students bring to the class discussion. Students who offer more than adequate responses may need learning extensions to expand their knowledge and challenge their thinking. Students who offer incorrect responses may need instructional support, which may take the form of additional collaborative work.

Observations happen when teachers circulate the classroom while students complete small-group or independent work during the Explore or Explain phases. Regardless of the structure, students should be engaged in the activity. If they are not, the teacher can diagnose the problem immediately with face-to-face interaction and offer encouragement and support as students work. For example, if a student seems to be isolating himself or herself from the small group, the teacher can join in and ask a probing question directly to the student in order to determine the cause of the lack of involvement. If a student is stuck on a written response during independent work, the teacher can ask clarifying questions to gauge whether the student is not responding because he or she lacks the content knowledge or because he or she lacks writing fluency. Each of these problems requires varying degrees of intervention. The teacher may intervene, allowing the student to participate in the group or complete the task rather than sit idle or submit a blank paper.

Observations are informal, but the feedback teachers gather from them is a valuable source of information. Teachers should take anecdotal notes regarding student actions when they happen and use these notes for deeper reflection at a later, more appropriate time. These notes can be useful in planning conferences with students and/or parents. Since the goal is to have students become scientifically literate, these notes act as roadside stops where the teacher can reflect on the student's progress and ensure that he or she remains on the right learning path.

## Student Work

Teachers can also ascertain whether students are "getting it" by reviewing the work they complete in response to learning at each phase of the 5Es. Students who list only one or two animals during a brainstorming Engage activity likely do not have extensive background knowledge to bring to a lesson on reptiles. Students who avoid using science equipment during Explore investigations may simply not know how to use it correctly. Students who turn in minimal or incomplete work or vague vocabulary responses during the Explain phase may not have the depth of understanding necessary yet. Perhaps students' response journals lack clarity and specificity. These are all signals that they need additional time, instruction, and involvement with the content to truly master the objectives. Seeing these inadequacies in work performance, teachers can plan additional and supportive instruction and collaborative work to build students' knowledge base and help them better meet the lesson objectives. Evaluating student work can also help determine where students need to cycle back through in the 5E process.

# Integrating Writing with Evaluate

During the Evaluate phase, teachers can have students write to demonstrate their understanding of content objectives. Written tasks are an important component of student assessment because they require students to summarize and synthesize their learning. Written tasks also provide students with real-life purposes for writing, ones where they may apply vocabulary, grammar skills, spelling, and punctuation in a meaningful context.

Many of the assessments discussed in this chapter included suggestions for students to use writing authentically. For example, in the troubleshooting problem-solving assessment in Figure 6.2, students were required to write to explain the process of photosynthesis. Writing becomes an important aspect of this assessment and is the means through which students demonstrate their final understanding of the learning objectives.

Additionally, the idea of having students write to demonstrate learning can be an integral formative assessment option throughout the phases of the 5E model. For example, Mrs. Pike, following her Explore simulation of natural selection (see Chapter 4), could have had students conceptualize the idea of natural selection in their science journals, then write one or two sentences to summarize what they had learned so far. An informal review of students' journal entries would provide a valuable formative measure of student learning of a very complex process. It is then the vehicle through which the teacher decides whether additional investigation is needed during the Explore phase or whether students are ready to move on to the Explain phase.

# Conclusion

Student assessment is an essential component of a teacher's learning plan. For an evaluation of learning to be complete, teachers should use a variety of both formative and summative assessment measures. Summative assessments include, but are not limited to, tests, quizzes, projects, performances, and problem-solving tasks. They are administered during the final phase of the 5E model—Evaluate. Formative assessments can take a variety of formats since they are an informal review of students' work and occur throughout the other phases. Together, they make up a complete and thorough evaluation of student mastery of the science concepts, information, and skills they are to have learned as they continue to grow and develop their scientific literacy.

## Questions for Reflection

1. Describe a time when you used formative assessment to guide or direct instruction. How did you determine that you needed to make a change? What did you change? How did it turn out?

2. Which type of summative assessment would you like to try with your students? What preparation do you need to do to implement this assessment?

# Differentiating Instruction in a 5E Classroom

The concept of differentiation might be second nature to some teachers. Yet others might lack sufficient experience with differentiation to understand it and implement it fully. Most teachers fall somewhere in between. Differentiation is important because not all students bring the same knowledge and skills to the classroom. Nor do all students have the same learning style or readiness level to adequately learn the science concepts their teachers need to teach them. Students learn at different rates. They learn through different instructional techniques. Because of these differences among students, teachers need to adjust the content itself, the process of delivering the content, and/or the assessment measures (products) they use to determine student mastery of the content (Tomlinson 1999). These adjustments are known as *differentiation strategies*. These strategies help to even the playing field so that all students can be successful with the content presented.

Carol Ann Tomlinson (1999, 11) identifies three areas where teachers may differentiate instruction:

1. content—what is taught

2. process—how it is taught

3. product—what students create

# Differentiating Content

Curriculum content is the "what" of teaching. When teachers differentiate content, they adjust the actual information or skills students learn. For many teachers, differentiating content is limited, depending on the structure and expectations set forth in their state or district's science curriculum standards. It is not the most effective method of differentiating in the science classroom, but it can be done during the Explore and Explain phases of the 5E model, if necessary. Here are some suggestions for differentiating content:

- Provide students with different investigation questions.

- Provide students with different resources to read and analyze.

# Differentiating Process

When teaching science, helping students become scientifically literate doesn't happen overnight. It doesn't happen by itself, except in the case of a very determined self-directed learner. Some students will take off running and try anything and everything on their quest for answers. Others will study facts and information, then make personal connections to the actual science. Others will seek out situations around them to make sense of information and bring meaning to it. Yet other students will need direct guidance and support to connect facts, information, scientific processes, and real-world applications. Teachers who differentiate process know and understand that one method of teaching science does not necessarily reach all learners. Therefore, they adjust and modify how they teach to meet the vast learning differences inherent in the students in their classes. The following are some suggestions for differentiating process:

- Allow students to listen to audio recordings of science texts.

- Match text levels to students' reading levels.

- Use flexible grouping for various learning activities.

- Have students restate the directions for their assignments.

- Provide word lists or semi-completed chapter summaries.

- Provide outlines or other organizers with pre-identified main ideas.

- Allow students to highlight, circle, and underline text to identify key words and main ideas.

- Provide students with teacher-created study guides.

- Use peer mentors and peer tutors to reinforce and support learning.

- Have students collaborate often.

- Use visuals (pictures, illustrations, charts, and models) from a variety of resources.

- Use kinesthetic activities to help students interact with their science learning.

- Use interactive online games and activities to stimulate thinking and learning.

- Allow students to lead and participate in discussions.

## Flexible Grouping

One way teachers differentiate process is through flexible grouping. This means that students work as a whole class in heterogeneous or homogeneous small groups, or with partners at various times throughout the 5E model. For example, Student A might begin a lesson by participating in a whole-class discussion (Engage phase). Next, Student A might work in a small, heterogeneous group with three other students (Students D, F, and L) to conduct an investigation (Explore phase). Then, he or she might work with a partner (Student G) to read about and summarize content (Explain phase). Following, Student A might work with a homogeneous group (Students E, L, and M) to complete a problem-based learning activity (Elaborate phase). Finally, the lesson might end with Student A working with his or her table partner again (Student G) to review before the final, summative assessment on a particular topic (Evaluate phase).

There are benefits to heterogeneously and homogeneously grouping students. Heterogeneous groups allow those students with less-honed skills to benefit from students who are in the know. Yet everyone can be active participants in group discussions and problem-solving tasks. Students can also work in homogeneous groups—that is, with students of similar ability levels. This grouping strategy would likely be used when students read a leveled text selection or complete a writing assignment. High-achieving students can collaborate with higher-leveled texts and complete open-ended written follow-up tasks while students with limited reading and writing abilities can work with the teacher to read a text with a lower readability level and complete a guided reading and writing task. When students collaborate with peers of the same ability, they challenge each other and support each other's strengths along the road to understanding scientific concepts.

# Differentiating Product

A third means of differentiating instruction is not in the instruction itself, but rather in the outcomes required of students. In most classrooms, students take summative tests to demonstrate their understanding of a particular concept. Indeed, the final phase of 5E is to evaluate students' level of mastery of the content. This phase of the learning cycle offers another opportunity for teachers to differentiate student work through student products. The following are some suggestions for differentiating student products:

- Modify/tier graphic organizers.
- Modify the type of recording sheets used to collect data.
- Modify/tier the types of investigation requirements students produce during the Explore phase.
- Allow students to choose the projects or investigations they complete during the Elaborate phase.

## More Differentiation Suggestions

Each phase of the 5E instructional model provides opportunities for each component of differentiation: content, process, and product. Different differentiation strategies fit better during different phases of 5E. This idea is illustrated in Figure 7.1.

**Figure 7.1** Differentiation Strategies for Each Phase of 5E

| Phase | Differentiation Strategy |
|---|---|
| Engage | • conduct paired conversations<br>• provide simple hands-on experiences<br>• allow students to draw ideas instead of writing them<br>• allow students to provide oral or written summaries<br>• allow students to make oral or written predictions |
| Explore | • allow students to work in homogeneous groups<br>• pair students homogeneously or heterogeneously<br>• assign group roles for investigations<br>• modify requirements for group lab sheets<br>• allow students to create demonstrations or models<br>• allow students to provide verbal explanations<br>• allow students to label and create pictures and illustrations instead of writing summaries<br>• create visual models to support each step of the investigation |
| Explain | • simplify vocabulary<br>• provide varied reading materials<br>• provide leveled reading materials and graphic organizers<br>• utilize small-group instruction<br>• embed definitions with vocabulary<br>• provide additional support for reading strategies and explanation of text structures<br>• utilize cloze activities<br>• provide thinking stems<br>• utilize leveled notes or outlines |
| Elaborate or Extend | • provide independent study projects<br>• allow for differentiated follow-up activities<br>• provide small-group, paired, or independent activities<br>• allow choices for activities that provide varying levels of complexity and hands-on opportunities<br>• allow for group summaries<br>• allow for varied presentation formats<br>• allow students to create demonstrations or models<br>• support use of verbal explanations instead of written responses<br>• allow students to respond with labeled pictures and illustrations<br>• provide visual models for each step of the Elaborate activity |
| Evaluate | • modify test items<br>• make assessment accommodations<br>• provide differentiated summative assessments<br>• utilize project-based activities<br>• utilize choices boards |

# Conclusion

The importance of differentiating instruction in a 5E classroom is second to none. Because students access content at various readiness levels and through different learning styles, differentiating instruction must not be overlooked. The strategies discussed in this chapter address the three areas of differentiation that Carol Ann Tomlinson identifies (1999): content, process, and product. These strategies provide teachers with the tools for differentiating their instruction so that all students can access content and demonstrate mastery of learning objectives. When teachers differentiate these aspects of their instruction, they provide students with varied opportunities to develop the skills necessary to be deemed scientifically literate in today's society. Instruction can be differentiated during any phase of the 5E instructional model, which allows students to experience and internalize their learning throughout an entire 5E lesson or unit.

## Questions for Reflection

1. Which component seems as if it would be the easiest to differentiate: content, process, or product? Why is this so?

2. Choose one of the differentiation suggestions provided in this chapter. How will you implement this suggestion in your next science lesson?

# Technology and the 5Es

Today's students demand and expect their teachers to provide them with instruction through technology. Up-to-the minute stats, data, and weather maps keep students informed with the most current information. Telecommunicating, blogging, Tweeting™, and live video feeds provide communication mechanisms for students to stay connected in this global learning environment. And application software (e.g., word processing, slideshow, and desktop publishing) and interactive online games, lessons, and learning options offer instructional support to teach students just about anything they have a mind to know.

Teachers in particular can take advantage of some of the amazing resources this digital age has to offer. This chapter offers highly engaging, meaningful, and resourceful integrated technology ideas as they pertain to each phase of the 5E learning cycle.

## Online Resources

Today's teachers can find just about anything and everything they need online to design, develop, and execute highly engaging and effective 5E lessons. For teachers lagging behind the 21st century technologies, conducting simple web searches usually yields substantial results. For example, a primary teacher wanting information and ideas related to soil can search for "soil information for kids." A middle school teacher wanting information related to the rock cycle can search for "rocks for sixth-grade students." A high school teacher who needs information on the carbon cycle can search for "carbon cycle high school." More specific topic searches will yield a greater number of appropriate site listings. But hunting for lesson ideas is just the beginning of a relationship with the World Wide Web. Online resources can fit into any phase of a 5E lesson design.

# Interactive Websites: Games, Activities, Simulations, and Quizzes

The use of an online interactive activity lends itself perfectly to the Engage and Elaborate phases of 5E. Teachers can find online matching games, vocabulary games, hangman games, and other games related to science topics. The Internet is filled with highly engaging, highly useful, and highly meaningful games and activities related to just about every science topic.

Teachers can also find useful simulations online. These are best suited for the Explore or Elaborate phases of the 5E model. Simulations are abstract realities. They model real-life situations (e.g., cause/effect relationships) in a fictional setting. To find a simulation, teachers can search "online science games" related to a topic of study. Many game searches will produce site options with simulations.

Finally, teachers can find online quizzes on just about any science topic at any grade level. These are best suited to use as review immediately preceding the Evaluate phase. Or teachers may opt to use them intermittently throughout the lesson plan to first pretest students' knowledge of the topic or concept, then provide periodic quick-checks to be sure students are grasping the information as the lesson continues. Online quizzes are usually about ten questions in length. Teachers can opt to have the class respond orally in unison and accept the most common response. Or teachers can have students collaborate with their table groups to respond. One group member can be responsible for recording the group's guess on a whiteboard. Teachers can also require students to write their answers on paper. Then, once the class choice is made, students have immediate, personal, and private feedback regarding their understanding of the topic. Keep in mind, however, that just because quizzes are available online, it does not mean that all of the questions are worthwhile to use with students. It is important for teachers to read and analyze the questions prior to using them with students to ensure their appropriateness for instruction.

## Smartphones, Tablets, and Apps

What would happen if students came to school and their teacher actually *encouraged* the use of their phones during class? Texting, Tweeting, Facebooking, and general gaming are some of the technology tasks students sneak into their daily routines, in spite of schools' attempts to disallow or discourage these actions during school hours. In the science classroom, these technology tools can be put to use to engage student thinking and compel their interest in classwork and learning. Smartphones and smartphone apps are technology tools teachers can embrace to support instruction in the classroom.

One use of smartphones and tablets is having students text or Tweet their teacher the answers to a short homework assignment. For example, teachers can assign students a reading selection for homework (or classwork), then require them to text or Tweet a summary to their teacher before the next day or class period. Students who do not have phones with texting or Tweeting capabilities should of course have the option to handwrite their summary on paper to submit to the teacher, or students might type it using a word processing program and print it or email it to their teacher. The advantage of texting/Tweeting is that the teacher knows students attended to their reading assignment. Students are limited in the number of characters they can use to text their teacher, so the assignment should not be overwhelming for them. However, summarization is a critical reading and writing skill that students should practice regularly, especially in the content areas (Marzano, Pickering, and Pollock 2001). And students can put their texting lingo to use for instructional purposes rather than exclusively for social means.

Another use of smartphones and tablets is their ability to provide instant access to information students need or answers to questions that arise during class discussions. This is accomplished simply by having students use their phone's web browser to search for and discover the answers to urgent and pressing questions. For example, when learning about the digestive system, what student wouldn't want to know how long the small intestine is compared to the large intestine? And how do the sizes of these organs compare between infants and grown adults? Any teacher can imagine these questions coming up naturally during the Explain phase of 5E. Yet the text resources the teacher is using may not answer these questions specifically. In a classroom where smartphone use is encouraged, students can collaborate (one group per phone/tablet) to research this information. Teachers can challenge students to be the first to find the answers to these (and other) questions, and write their findings on whiteboards. After students have discovered their answers, the class can compare notes. This would make a great topic of conversation if the groups discover different answers. A follow-up homework assignment could be to have students theorize as to why the data was not the same in each group. This would require students to record data from each group, analyze the totals (find the mean and range), and then think and respond critically to explain why the results may have differed. This type of instant learning opportunity would not be possible without smartphones or tablets. In a traditional classroom, the teacher may have asked one or two students to find and bring the answer to class the next day. Instead of involving just one or two students, the whole class gets involved in this mini-research project, and students are offered the opportunity to apply critical-thinking skills during the process of discovery learning.

Finally, teachers can allow students to download science apps and then access and use them as part of a well-designed and well-developed instructional plan. Depending on content and implementation, these types of apps will most likely fit into any of the first four phases of the 5E instructional model. Science-related apps can also provide the means for students to engage in meaningful, real-world, and relevant homework assignments. However, when doing so, teachers will need to provide alternative yet equally relevant options for students without smartphones or the ability to download and use apps.

## Online Independent Research

Researching online is an important skill for students to master in order to be considered college and career ready. This can easily be integrated into the Explain or Elaborate phases of the 5E model. Independent projects generally translate into research papers; however, students can use online resources to compile and summarize information related to a topic of interest to suit any writing format. For example, elementary students can research the skeletal structure of a secret animal, create a three-dimensional model of the skeletal structure using twist-ties and/or pipe cleaners, then write clues about their animal so that other students can guess what it is. Middle and high school students can design and create a planetarium to illustrate and explain newly discovered planetary systems. To accompany their planetarium, students can write monologues to describe these new systems and explain how knowledge of our own solar system applies to these newly discovered systems. In both cases, students use Internet resources to conduct their research and then compile their research into an interesting, creative, and original summary.

One consideration teachers should have regarding students' use of the Internet is their age and computer abilities. The younger the student, the more scaffolding, support, and collaboration they will likely need to succeed with their independent project. All the same, middle school teachers should not simply assume students know how to conduct independent research projects. Some students in class may not have the skills to succeed completely on their own. Also, they may lack clarity regarding the teacher's expectations for a completed project. Teachers can address these concerns by surveying students about their past experiences with research projects, allowing students to pair up to conduct their research, and providing clear and compelling rubrics and completed models for students to use as a guide for acceptable work.

The complexity of the task is also an important independent research consideration. Students who are just starting to conduct online research would likely benefit from teacher-directed lessons, with follow-up work that requires fill-in-the-blank or short-answer responses to guiding questions. As students mature, teachers can feel more confident assigning tasks to them with less direction. A summary sheet with open-ended guiding questions might be more suitable for students at this level. Eventually, teachers can simply assign a research project by setting students to their task and defining the project outcomes. These open-ended projects require little teacher supervision save for periodic check-ins to be sure students are on track to complete their project in the time allotted. Although teachers can allow students to assume more and more responsibility with regard to independent research projects as they climb through the grade levels, some students may still need support. It is important that teachers stay in tune with their students' abilities to effectively utilize the Internet in this way.

Teachers can also use independent online research projects to teach students about reliable and appropriate Internet websites and how to conduct responsible online research. Some ideas to consider are listed on the following page. Science teachers can attend to this instruction themselves. Or they can elicit the assistance of a technology teacher, a school media specialist, or a public librarian to conduct a lesson on the safe and effective use of the Internet for research purposes.

## Instructional Considerations for Responsible Online Research

Teach students how to…

- evaluate the information they read for accuracy and correctness

- identify information from respected and credible authors and sources

- distinguish online information as fact, fiction, or opinion

- conduct efficient and effective online searches

- cite information appropriately and respect copyrights

## Published Instructional Resources

Many textbook companies provide online components for both teachers and students. A local school district may require its teachers to use one particular published program or series of programs. This limits the outside resources they can bring into their classrooms. Fortunately, these companies have built-in technology components. Teachers and students need to simply access them. For example, textbook pages may include links to websites for students to watch short videos or read additional information about a particular topic. Students may even be able to access the textbook itself online. They can find games, interactive activities, and quizzes through online links that are directly connected to the texts they use. Teachers, too, have online resources available to them to use as part of a comprehensive study of science. Some texts even come with ready-made tests and quizzes that interface with student response systems. (These are "clicker" systems that students use to electronically submit their answers to multiple-choice tests in an electronic database.) Here, the electronic program scores students' responses and organizes this data for teachers. No more test grading! The student response system does all the work. Teachers whose school district requires the exclusive use of a core science program should be able to find and use online resources to directly engage students and provide them with instructional support as they continue along their journey toward scientific literacy.

In addition to online textbook support, some favorite science-related trade books, such as *The Magic School Bus* series, and informational favorites, such as *TIME® for Kids,* have accompanying online games, activities, quizzes, videos, and lesson resources for both students and teachers. These resources are especially useful during the Explain phase. They also provide adequate supplemental reading and viewing material for students who wish to extend their own learning during the Elaborate phase.

## Virtual Experiences

Some schools may not have the funding available to support the purchase of science tools (e.g., microscopes, balances, thermometers, Bunsen burners) or supplies. However, as long as teachers have access to a computer and projector, the entire virtual world is open to them and their students. No money for owl pellets? No problem! The Internet has online virtual owl pellet dissection websites, complete with interactive bone separation, bone charts, and reproducible student activity sheets. No money in the budget for ammonia, iodine, litmus paper, or agar? No problem! Online virtual chemistry labs have all these materials, and they provide essential digital lab equipment, such as beakers, scales, and pH meters. Don't have the means to travel around the world for soil samples? That's okay, too. Students can find online access to sites with various soil samples. They can observe the basic types of soil, learn about the makeup of the soil, and make connections between the characteristics of the soil and its location. So even though personal hands-on experiences are the preferred method of "doing" science, today's online resources provide an endless supply of virtual investigation possibilities.

Likewise, some information is best shared through experts in their fields. With today's technology, teachers can use video chat technology, blogs, email, and live or pre-recorded and archived informational sessions to bring experts right to the classroom. Universities, television networks, and social networking sites are often great resources to find these experts. There may be live video chatting options, blogs, or other communications postings either from the news channel website or from a Twitter or Facebook page. If a teacher is determined to find an expert in a particular field or area of study, chances are he or she will find just the right person to bring to the classroom through digital means.

Finally, online resources provide ample full-length movies, short videos, and video demonstrations for classroom use to help students experience science virtually. To find a video of choice, simply search for the science topic along with the words "videos for kids." Searches will likely yield a multitude of appropriate, pertinent results. Teachers will not need to take much time to preview and find just the right video for their students.

The ideas in Figure 8.1 summarize just some of the options available to teachers for using online resources throughout all phases of a 5E lesson plan. See Appendix A for additional suggestions.

**Figure 8.1** Online Resources for Each Phase of 5E

| 5E Phase | Online Resources |
|---|---|
| Engage | • games<br>• activities<br>• virtual experiences |
| Explore | • simulations<br>• virtual experiences |
| Explain | • texting or Tweeting summaries<br>• searches using smartphones, tablets, or web browsers<br>• online resources that accompany core instructional programs (textbooks)<br>• virtual experiences |
| Elaborate or Extend | • games<br>• activities<br>• simulations<br>• independent research projects<br>• virtual experiences |
| Evaluate | • online quizzes<br>• independent research projects |

# Application Software

The previous section of this chapter discussed the use of online applications within the construct of the 5E model. This is not the only effective means of using technology in science class. Teachers and students can use application software to complete engaging and meaningful evidence of learning. This may take place while students are learning about a concept (during the Engage, Explore, Explain, and Elaborate phases), or as a summative assessment of learning (during the Evaluate phase).

Application software is any computer software program that allows users to document, record, organize, or share information. Word processing and information database software are two examples. Other application software may be specific to a particular task. This software works with other digital tools and allows students to extend the senses electronically. Digital microscopes and digital thermometers that interface with computer software are also examples of this. Students can use the digital equipment, then download, organize, and display their data through a software program on their computer. This idea also applies to digital cameras and video recording equipment, which students might use to document evidence with regard to an investigation or research project.

The uses of application software as they relate to science learning are endless. The summary in Figure 8.2 is only a sampling of the ideas that come to mind when visualizing the use of application software as a means to learning.

**Figure 8.2** A Short List of Software Applications in Science

| 5E Phase(s) | Software Application | Sample Uses |
|---|---|---|
| Explain<br>Elaborate | word processing | • summaries<br>• vocabulary cards<br>• concept sorting cards<br>• notes<br>• graphic organizers<br>• concept maps<br>• visually summarizing webs<br>• hierarchies, chains, cycles |
| Explore<br>Explain<br>Elaborate<br>Evaluate | desktop publishing | • calendars (to record data or observations over time)<br>• fliers or posters displaying important information<br>• how-to or informational brochures |
| Explain<br>Elaborate<br>Evaluate | presentation software | • slideshow presentations (summaries, explanations, procedures)<br>• videos or movies of an investigation<br>• podcast summaries of information |
| Explore | database software | • data collection and organization<br>• graphing applications |
| Explore<br>Elaborate<br>Evaluate | digital equipment | • digital cameras for documenting steps to an investigation, recording results, and sequencing events<br>• video cameras for making movies about investigations, observations, or research projects<br>• digital microscopes for magnifying objects<br>• graphing calculators for taking, recording, and displaying data |

# Using Technology in Classroom Settings

Some schools are fortunate to have one computer for every student. Others are lucky just to have one computer per classroom. Regardless of students' access to technology resources, teachers should remember that these are simply the resources to *support* and *extend* learning for all students throughout the phases of the 5E model. With so many options available, teachers may feel overwhelmed to the point where they don't use anything. Or they may feel obligated to use everything. The former situation is not acceptable. The latter situation is simply not feasible. As long as teachers set clear learning objectives for their students, the technology resources they find or want to use will naturally fall into their proper place throughout the phases of the 5E model. The technology is there to support teachers with regard to their instructional mission; it is there to support students along their path to becoming scientifically literate. The ideas that follow suggest how teachers may accomplish their objectives with digital support.

## Whole-Class and Small-Group Technology Uses

Ideally, every student (or small groups of students, in the case of shared work) has access to online information, digital equipment, and computer software to begin, develop, and finalize a complete project or assignment. With the onset of tablets in classrooms as a means of replacing textbooks, individual computer access may become a staple in public schools. However, in some classrooms today, that is simply not the case. For classrooms without the benefit of equipment for every student, the ideas set forth in this chapter are still manageable. The classroom setting will just look different from those with full access.

Of course, students should be the ones to engage with the computer whenever possible to access information, an activity, or an investigation. When this is not possible, teachers will still want to do what they can to create access to these resources for their students. Additionally, some activities are best suited for whole-class delivery. This may be when a teacher is first kicking off a lesson with the whole class, or when the whole class can work collaboratively to process a particularly difficult concept. Using single computers to share instructional content can work through many organizational means, such as the following:

- The teacher uses a computer or document camera and projector to display the website or information with the whole class. The teacher can facilitate the class discussion but still have students converse in pairs or small groups to make observations, discuss content, or summarize information.

- The teacher uses a computer, a projector, and a wall or an interactive whiteboard to display an interactive game, activity, or simulation. The teacher randomly calls on students to participate at the computer or the board. Additionally, students can work in groups to strategize, then send one representative to the computer or board to interact with it.

- The teacher sets students to an online task in the classroom, then utilizes a computer lab for students to work individually or in pairs to complete the assignment.

- The teacher demonstrates a computer-based activity for the whole class and then divides students into groups. One at a time, each group completes the online or interactive activity at computer stations while other groups are engaged in noncomputer-based activities (e.g., reading, writing, hands-on activity or investigation). After a set amount of time, students who were at the computers rotate to an alternative activity, and the next group of students completes the computer-based activity. This rotation system continues until all students have had a turn at the computers.

## Individual Technology Uses

For classrooms where each student has ready access to his or her own computer or tablet, the objective for the teacher becomes more a situation of balance between independent computer work and actual hands-on investigations and collaborative problem solving. Complexities regarding the organizational systems, classroom management strategies, and accountability aspects of the lessons remain the same, although the methods and structure in these classrooms may differ from classrooms with just a few computers. The following are a few ideas regarding the use of individual computers or tablets in the classroom:

- Students can use a word processing program to type or write notes and summaries about their thoughts or predictions during the Engage phase, investigations they may have conducted during the Explore phase, or information presented during the Explain phase.

- Students can access informational text and resources from educational publishers online during the Explain or Elaborate phase.

- Students can use application software to reflect, summarize, predict, synthesize, document, or demonstrate understanding of the concepts presented during the Explain, Elaborate, or Evaluate phases.

# Conclusion

Technology is a fact of life for teachers and students in the 21st century. Instructional institutions have an obligation to maximize the utilization of technology as much as possible, but only as a thoughtful, meaningful tool to help students learn the content, concepts, and processes that encompass all aspects of scientific literacy. Fortunately, organizations, companies, and individuals have created numerous relevant, engaging, and informative products and resources for teachers to use during all phases of the 5E model. There are a lot to choose from. However, for teachers who plan well and maintain clear learning objectives for their students, technology becomes a welcome tool to purposefully use within the 5E instructional model.

## Questions for Reflection

1. Which online resource(s) will you use to support your students' science learning? In which phase will you utilize this resource? How will it support your students in the learning process? (What are your objectives for using this resource?)

2. What application software do your students have access to that will support their science learning? How will you integrate this during your science lessons?

3. What challenges do you believe interfere with your students' ability to utilize technology as an integral part of their science learning? How can you overcome these obstacles to maximize the technology use in your classroom?

# Chapter 9

# Putting It All Together

Just as students come to class with varying levels of background information, abilities, and aptitudes to learn new things, teachers, too, have a range of talents and experiences along with their preferred teaching styles that they use to teach new ideas. This includes the implementation of the 5E instructional model of teaching science. Teachers who regularly use small groups, hands-on activities, and exploration in their classroom will likely need less support in implementing this instructional model than teachers who have little or no experience with this style of teaching. Regardless, the suggestions and steps in this chapter will help every teacher implement the 5E model with success.

## Steps to Success

In Chapter 1, readers were introduced to Robert Marzano and Debra Pickering's (2011) research related to student engagement. One question students ask themselves when deciding whether or not to become engaged with a topic is, "Can I do this?" (15). Teachers may be asking themselves the same question! The answer, simply, is "Yes!" The 5E instructional model provides a structure around which teachers may build instructional experiences that lead students to greater scientific literacy. The structure of the 5E model is organized the way it is for a reason. Ideally, this model would be implemented wholly for each science lesson. However, some teachers may be overwhelmed at the prospect of restructuring their entire science program. These teachers may be more comfortable trying one or two phases in the series before implementing the model completely. Regardless of whether teachers jump in fully or just begin by dipping their toes, the suggestions that follow will help them be successful with implementing the 5Es in their science classrooms.

1. Set clear learning objectives.

2. Establish clear and comprehensive evaluation criteria.

3. Design effective learning activities.

4. Establish clear expectations and classroom routines.

## Step 1: Set Clear Learning Objectives

When some people go on vacation, they simply jump in the car and go. Their destination is unclear, and their path is not predetermined. They may have some idea of their overall direction of travel, but the specific outcome is unknown when they start out. Although this is an acceptable plan for summer trips, in the classroom, this open-ended approach to instruction simply does not work. Learning objectives determine the destination. The teacher's instructional activities provide the learning path for getting there. The first, most important step to planning an effective 5E lesson is setting a clear destination. This is accomplished by writing learning objectives. These are likely determined by the curriculum standards. To accomplish this, teachers simply need to rewrite the standards as objectives. This is done by analyzing the standard and writing out exactly what students should be able to do as a result of learning that standard. Examples of how this might look are included in Figure 9.1. For the sake of consistency, the remaining suggestions in this chapter reflect back on these objectives to illustrate the process of curriculum design following the 5E instructional model.

**Figure 9.1** Physical Science Content Standards Rewritten as Learning Objectives

| Grade Levels | Curriculum Standards | Learning Objectives |
|---|---|---|
| K–4 | Understands the properties of objects and materials | Students will observe and measure objects in terms of their properties, including size, shape, color, temperature, weight, texture, sinking or floating in water, and reaction and repulsion of magnets. |
| 5–8 | Understands properties and changes of properties in matter | Students will classify and compare substances on the basis of characteristic properties that can be demonstrated or measured (e.g., density, thermal or electrical conductivity, solubility, magnetic properties, melting and boiling points) and know that these properties are independent of the amount of the sample. |
| 9–12 | Understands structures and properties of matter; chemical reactions | Students will distinguish between physical and chemical properties of matter and physical and chemical changes of matter. |

## Step 2: Establish Clear and Comprehensive Evaluation Criteria

Once teachers have a clear destination (learning objectives), they should establish clear, comprehensive assessment criteria upon which to evaluate student learning—the Evaluate phase. The rationale for this backward-planning approach is that teachers who now plan how they will know if students have reached the desired results (via evidence of assessment) will ensure that "the course is not just content to be covered or a series of learning objectives" (Wiggins and McTighe 1998, 12).

Ideas related to using methods that effectively determine mastery of the content objectives were discussed at length in Chapter 7. Ideally, teachers will organize a balance of tests, quizzes, performance tasks, and/or projects for students to complete to unequivocally demonstrate their understanding of the content objectives. The lesson examples in Figures 9.2–9.4 demonstrate how teachers might formatively and summatively evaluate student learning for each of the physical science objectives listed in Figure 9.1.

## Step 3: Design Effective Learning Activities

Now that the final destination is set and the evaluation criteria are established, teachers can plan the effective learning activities for each of the first four phases of the 5E model that will make up the "path" to content mastery. Learning activities should be matched to the learning objectives. For example, teachers using the learning objective for grades 5–8 in Figure 9.1 would not plan a review of the periodic table of elements since the learning objectives clearly require students to review the properties of solutions rather than their chemical structure. Although these ideas are related, activities specifically related to the periodic table of elements would have a different set of objectives. If knowledge of the periodic table of elements is a prerequisite for these objectives, the teacher would have conducted a lesson with a related objective prior to developing this lesson on comparing properties.

When designing effective learning activities, teachers should keep their students in mind. Every class has its own culture and climate. In some classes, students can work collaboratively with little direction. In others, collaborative activities will lead to conflict and unhealthy competition. In the former class, the teacher can organize creative, open-ended, meaningful investigations mostly run by students. In the latter class, the teacher will want to organize more structured, teacher-centered activities and investigations until students have learned how to work together more effectively.

Additionally, a kindergarten teacher's approach to classroom investigations will look very different from a fifth-grade teacher's approach. Kindergarten students are still learning how to work independently and cooperatively. They need constant supervision, positive reinforcement, and shorter work times. Fifth-graders should be able to handle a teacher giving a set of directions, setting the expectations, and letting them get to work. All students can benefit from experiences that lead them to becoming better collaborators and communicators. These experiences will just need to be designed differently to better match the cognitive capabilities and maturity of the students in the class.

Teachers should also remember that the more active the students are as participants, the greater their commitment and attention to the topic about which they are learning. Engaging work does not always equate to hands-on activities, although this is an important component of engagement. Students can be more active learners when reading from textbooks, when their teacher allows them to read information with a partner, when watching informational videos, and when their teacher provides them with outlines or graphic organizers to complete related to the content. Increasing students' level of activity gives their learning meaning and purpose, and it holds students accountable for their learning. The lesson samples in Figures 9.2–9.4 outline a 5E lesson for each of the learning objectives listed in Figure 9.1. Although the learning activities collectively demonstrate the structure of a 5E lesson, teachers should remember that not every activity will work in every classroom and that they need to consider their class's culture and climate, and their students' abilities, background knowledge, and maturity levels.

## Step 4: Establish Clear Expectations and Classroom Routines

Teachers who establish clear expectations have students who know what is expected of them and who can strive to meet the objectives set forth by their teacher. The idea of setting clear expectations was introduced in Chapter 4 as a classroom management tip for organizing classroom investigations. The same is true when organizing effective 5E lessons. A teacher who sets clear learning objectives throughout all phases of a 5E lesson can better maintain a structured, organized, and effective learning environment for students.

# Keep It Simple!

Teachers will likely come across a multitude of activities to support any number of learning objectives as part of their science coursework. It is easy to become swept up in this idea and that idea. Some teachers will want to do it all! Unfortunately, time restrictions will limit the types and lengths of activities teachers can use as a regular part of their instructional plan. And using too many activities can lead a lesson astray, causing the learning objectives to become murky and off topic. Throughout Chapters 3–7, teachers were encouraged to carefully reflect on the learning objectives of the activities that were chosen. Clear objectives lead to clear and focused lesson design. Teachers who maintain a steady path toward the defined destination will do so by planning simple yet effective activities. Keeping it simple will keep the lesson on course and maximize students' potential to master the objectives and contribute to their scientific literacy.

# 5E Lesson Samples

The 5E lesson plans that follow in Figures 9.2–9.4 were developed to support the learning objectives in Figure 9.1. These lesson ideas are simplified here for the sake of space and for the flexibility of classroom implementation. For example, in the Explain phases, the lesson plans list resources by category (e.g., textbook, trade book, article), not by specific title. This was done intentionally so teachers may utilize the text resources available to them and their students. The grades 9–12 Elaborate activity mentions an online simulation, but it does not specify a particular one so as not to limit the types of web-based activities used in the classroom. Teachers should feel as though they can use the resources available to them that suit their own teaching styles, their students' learning styles, and allow for flexibility to meet the needs of their students. Rest assured, all of the resources listed do exist in one way, shape, or form. For a 5E Lesson Plan Outline, see Appendix C.

**Figure 9.2** Sample 5E Lesson for Grades K–4

| Objective | Students observe and measure objects in terms of their properties, including size, shape, color, temperature, weight, texture, sinking or floating in water, and reaction and repulsion of magnets. |
|---|---|
| Engage | Activity: Students sort objects into groups and justify their categories. |
| | Formative Assessment Opportunity: recorded list of two categories |
| Explore | Activity: Students choose four objects from the Engage activity to observe, measure, and record properties on a matrix. |
| | Formative Assessment Opportunity: matrix |
| Explain | Activity: Students read textbooks, guided readers, or trade books. |
| | Formative Assessment Opportunity: mini-book of objects and their properties |
| Elaborate | Activity: Students use their senses and equipment (balance scales and magnets) to predict what is inside a closed box. |
| | Formative Assessment Opportunity: science journal reflection |
| Evaluate | Activity: Students label a picture summary and take a property matching quiz. |
| | Summative Assessment Opportunity: completed summary and quiz |

**Figure 9.3** Sample 5E Lesson for Grades 5–8

| | |
|---|---|
| Objective | Students classify and compare substances on the basis of characteristic properties that can be demonstrated or measured (i.e., density, thermal or electrical conductivity, solubility, magnetic properties, melting and boiling points), and know that these properties are independent of the amount of the sample. |
| Engage | Activity: Ask, "What do water, catsup, oil, candle wax, and ice pops have in common?" Students respond with ideas on sticky notes and place them on an anchor chart. |
| | Formative Assessment Opportunity: sticky note anchor chart |
| Explore | Activity: Students participate in the following activities: salt water density activity (small groups); "Conducting Solutions" lab investigation (whole class); melting and boiling point comparison chart (pairs); dissolving candy in different colored water experiment (small groups); magnet check of all substances. |
| | Formative Assessment Opportunity: lab investigations, science journal reflections, comparison chart |
| Explain | Activity: Students read textbooks, online articles, and other texts. |
| | Formative Assessment Opportunity: text summaries, science journal reflections, vocabulary maps |
| Elaborate | Activity: Students play a solubility interactive game (whole class). |
| | Formative Assessment Opportunity: recorded results in science notebooks |
| Evaluate | Activity: Students write a summary of the Engage question related to characteristic properties. They also take a quiz and write a short story titled, "A Day in the Life of a Solution." |
| | Summative Assessment Opportunity: summary, quiz, story |

**Figure 9.4** Sample 5E Lesson for Grades 9–12

| Objective | Students distinguish between physical and chemical properties of matter and physical and chemical changes of matter. |
|---|---|
| Engage | Activity: Ask, "Is every substance soluble?" Students respond with ideas on sticky notes and place them on a chart. |
| | Formative Assessment Opportunity: sticky note chart |
| Explore | Activity: Students participate in a lab: Physical and Chemical Changes (small groups). |
| | Formative Assessment Opportunity: lab record sheet, science journal reflection |
| Explain | Activity: Students read textbooks, online articles, and summaries, and view an online video. |
| | Formative Assessment Opportunity: categorizing worksheet and notes, vocabulary maps, comprehension questions |
| Elaborate | Activity: Students complete an online simulation on phase changes. |
| | Formative Assessment Opportunity: simulation summary |
| Evaluate | Activity: Students take a final test. |
| | Summative Assessment Opportunity: final test |

# Classroom Organizational Strategies

The National Science Education Standards (1996) are centered on the premise that learning science is an "active process" (2). In addition to learning content, students must think critically about the world around them. They should also learn and apply process skills, such as observing, inferring, experimenting, questioning, and communicating. For students to be truly active learners in the science classroom, teachers must provide an appropriate learning environment. Teachers can foster this creative learning environment throughout the phases of the 5Es by organizing both the physical and social aspects of their classrooms.

# The Physical Environment

The 5E instructional model is a framework for lesson design. Embedded within this design are elements of effective instructional practices, such as the use of paired discussions and small-group work. Teachers whose classrooms mirror a traditional classroom setting with desks in rows and each desk acting as its own little learning island will need a plan to restructure the physical environment to allow for paired and small-group work. If students will be moving their desks during class to form groups, the teacher will need to allow time for students to form their groups before the activity and return their desks to their original positions after the activity. Teachers will also need to determine who will work together as part of their instructional plan. Teachers may have students form their own groups, design a strategy for groups to be randomly created, or dictate the group makeup through some means.

Aside from the physical placement of students, teachers who effectively implement the 5E instructional model will also need to consider how they organize equipment and materials. This idea was first introduced in Chapter 4 as a consideration for conducting successful investigations. Equipment and lab materials are only one part of the 5E model to consider. Teachers also use text resources, online resources, handouts, notebooks, posters, charts, and displays, so it is important to plan for the use, storage, and display of each of these at the point in the lesson when they are needed.

Sometimes, teachers will want to involve students in kinesthetic activities. During these times, it is important to consider how students will move about the classroom. However, in classrooms with limited space, the teacher may want to organize the routine so that the materials or equipment move from group to group rather than the groups moving from one set location to the next. For example, Mr. Burns, during the Explain phase, wants to have students collaborate to summarize information about animal defenses. He prepares five posters, one labeled with each of five animal defenses: camouflage, shape changes, movement, trickery, and chemicals. Each group will have just two minutes to record everything they can about each particular animal defense. However, his classroom space is limited. It does not facilitate movement of groups of students from chart to chart. Instead, he organizes the students into groups. Each group gets a different color marker. He gives each group one poster to start. When time is up, the posters rotate to the next group. At the end of the review, each group receives the poster with which it started, and then record any final ideas not added from the other groups. One spokesperson reviews the content from all the groups to the whole class, and the posters are posted in the classroom for everyone to see.

Both the physical structure of furniture and supplies and the organization of student groups will lead into time management considerations. Teachers who think ahead and plan for physical movement of desks, students, materials, and equipment will minimize class time spent preparing for collaborative activities.

## The Social Environment

The most important aspect of any successful classroom is a positive teacher-student relationship. According to Robert Marzano (2007), this involves the teacher creating a climate of teacher guidance and control and a climate of teamwork. Consideration to the physical climate, as discussed in the previous section, leads to at least the appearance of an organized, in-control classroom setting. Teachers should also have a well-established set of expectations with clear consequences that fairly apply to all students. In addition, teachers can establish their position as the classroom leader by welcoming each student to class, using students' names, smiling, and using humor when appropriate. If students perceive that their teacher is in control of all aspects of the classroom environment, they will usually respond respectfully toward the teacher and the work he or she requires of them.

Equally as important as the teacher's interaction with his or her students is the teacher's expectation for collaboration and teamwork. This can be established, as shared in Chapter 4, through setting clear expectations for interactions among students when they are involved in paired or small-group work. Likewise, the teacher must communicate through both verbal and nonverbal means his or her support of respectful interaction, even when there is disagreement. Students who feel that they are valued both individually and as part of their classroom community will strive for greater success in the classroom.

# Making Time for Teaching and Learning

Time is one valuable resource over which classroom teachers have little control. A multitude of other curriculum demands and a finite amount of instructional time devoted exclusively to science limit what teachers can accomplish with their students. However, teachers can manage their time to maximize the time they spend engaging students in meaningful work, thus leading students to greater success with content.

## Time Management Suggestions

These short-range time management suggestions reveal how teachers can maximize the time with which they engage students in science content from day to day.

As previously mentioned, time management is sometimes all about organization. The more organized a teacher is, the less time he or she takes to gather, distribute, and recollect materials. So the first strategy for maximizing class time is to **get and stay organized**. Any lab demonstrations should be completely set up and ready to go at the start of the class period. This will ensure an efficient transition from lesson to activity and back again.

Additionally, teachers should **plan and visualize how materials will be distributed and re-collected** as part of their overall lesson plan. Depending on the age of students, the materials being used, and classroom space constraints, teachers can distribute and re-collect materials themselves, assign one member from each pair or group to collect and return materials to a specific location, or assign class helpers to conduct this task for all students and groups. Regardless of the structure, the delivery and return of classroom materials and equipment should be quick and nonintrusive to the students' work time.

Teachers can **put technology to work** to lessen time limits for classroom activities. They can download and use a digital stopwatch or timer to hold both themselves and the students accountable for their own time management. For example, Mrs. Quail frequently likes to give students "talk time" with partners. But this can sometimes intrude on her instructional time. She may lose track of how much talk time she gives her students, or they may begin asking questions related to but off-topic from that day's lesson. To address the former issue, she instructs students to discuss just one topic or try to answer just one question for every eight to ten minutes of teacher talk time. She poses her question, then sets a digital timer for 90 seconds. Once it rings, she continues with her instruction. To address the latter issue, she has a ready supply of sticky notes at each table. As questions arise during her instruction or talk time, students make note of their questions. Mrs. Quail collects the students' questions at the end of the lesson to review before the start of the next day's class period. Questions she can answer she does at the start of the next class period. Questions she cannot answer or does not have time to answer she posts to a bulletin board and invites students to take a question, research it, and answer it on the classroom blog. Teachers who do not have access to or cannot use digital timers can use everyday kitchen timers for this same purpose.

While working, students can frequently lose sight of their task objectives. The **use of timers or stopwatches during student work time** can help them stay aware of the remaining time they have to complete a specific task. Additionally, posting a "to do" checklist of tasks to complete can help students ensure that they finish everything necessary before time runs out.

Finally, teachers who **stay focused on the lesson objectives** can minimize "off task" time in their classrooms. Younger students in particular have a tendency to share whatever idea comes to mind, unrelated as it may be to the science lesson the teacher is conducting. For example, when Mr. Drake has his first-grade students seated on the floor at the front of the room for a shared science reading, he knows that Vince will inevitably raise his hand and share a story or idea about...whatever: frogs, ice cream, his pencil, kickball, etc. Mr. Drake has learned to validate Vince's statements, then immediately redirect the class to the information before them. He is also in the habit of saying, "That is a great story about _____. Let's think about how we can use that story to explain how _____ occurs" when any student is determined to drag the class off on an unrelated tangent.

## Lesson Planning Checklist

As teachers begin to implement the 5E instrumental model to its fullest extent, they can monitor their own goals and objectives by reflecting on their personal instructional practices and evaluating their progress. The first step to implementing the 5Es is planning 5E lessons. The 5E Lesson Plan Outline in Appendix C may help with this. As teachers continue to develop 5E lessons and implement them with their students, they can refer to Figure 9.5 to ensure that they are on track to maintain a steady course with the 5E model as the driving force of their instruction.

**Figure 9.5** Lesson Planning Rubric

| | | | |
|---|---|---|---|
| My lessons include clear and concise learning objectives. | Always | Sometimes | Never |
| My lessons follow the 5E model. | Always | Sometimes | Never |
| I *engage* student thinking before starting a new topic. | Always | Sometimes | Never |
| Students *explore* ideas with a hands-on or minds-on activity before they begin to learn facts and information about a specific topic. | Always | Sometimes | Never |
| I use leveled and varied texts and informational resources to help *explain* topics and concepts to students. | Always | Sometimes | Never |
| I use activities that *elaborate* on the informational learning and allow students to *extend* their thinking. | Always | Sometimes | Never |
| I use a variety of both formative and summative assessments to *evaluate* student learning. | Always | Sometimes | Never |
| The activities in my lessons directly connect to the learning objectives. | Always | Sometimes | Never |
| Students are *engaged* in whole-class, small-group, paired, and independent tasks on a weekly basis. | Always | Sometimes | Never |
| My students know what they are learning and why they are learning it. | Always | Sometimes | Never |
| My classroom is organized, and I have materials ready for the daily activities. | Always | Sometimes | Never |
| My students write frequently as a natural part of the learning process. | Always | Sometimes | Never |

# Conclusion

This chapter focused on the overall implementation of the 5E instructional model as an effective method for planning science instruction. Each teacher and classroom is unique; there is no "one size fits all" when it comes to science instruction. The 5E instructional model inherently supports this diversity from grade to grade, from class to class, and from school to school. Teachers can use the lesson plan outline to structure lessons with a greater or lesser emphasis on hands-on activity, informational reading, or independent learning, depending on the abilities and aptitudes of their students. The reflection questions that follow allow teachers to analyze where they are along the path to implementating a 5E lesson. Teachers who implement the 5E instructional model can look forward to exciting, meaningful, relevant, and engaging work as students continue along their lifelong paths to scientific literacy.

## Questions for Reflection

1.  On a scale of one to five, rate how well your lesson plans fit the 5E instructional model *before* reading this book. Justify your rating.

2.  Let's assume your goal is to reach a five on this same scale. What is your target date to reach this level?

3.  What do you believe is the easiest phase to implement? Plan and deliver an activity for an upcoming lesson for this phase. Explain why you were or were not satisfied with this activity. To do it again, what would you change, if anything?

4.  What do you believe is the most challenging phase to implement? Plan and deliver an activity for an upcoming lesson for this phase. Explain why you were or were not satisfied with this activity. To do it again, what would you change, if anything?

5.  When you're ready, plan and execute a complete 5E lesson. Reflect on its effectiveness. Celebrate your successes! Write about your proudest moment. What changes might you consider for next time?

# Recommended Resources

## Books

Carle, Eric. 1987. *A House for Hermit Crab*. New York: Aladdin Paperbacks.

——. 1994. *The Very Hungry Caterpillar*. New York: Philomel Books.

Cole, Joanna. 1986–2012. *The Magic School Bus* series. New York: Scholastic.

Hoban, Russell. 1969. *Bread and Jam for Frances*. New York: Harper Trophy.

O'Dell, Scott. 1984. *Island of the Blue Dolphins*. New York: Dell.

## Digital Resources

BBC Science Clips
  http://www.bbc.co.uk/schools/scienceclips/index_flash.shtml

BrainPOP®
  http://www.brainpop.com

BrainPOP® Jr.
  http://www.brainpopjr.com/

ChemCollective
  http://chemcollective.org/teachers/index

CSI: Web Adventures
  http://forensics.rice.edu/

Discovery Channel™ Videos
  http://dsc.discovery.com/videos/

Discovery Educator Network
  http://blog.discoveryeducation.com/

Explore Learning "Gizmos"
   https://www.explorelearning.com/

The Futures Channel©
   http://www.thefutureschannel.com/

Jefferson Lab
   http://education.jlab.org/indexpages/elementgames.php

Just Kids Games
   http://justkidsgames.com/

KidWings
   http://www.kidwings.com/

Learning Games for Kids: Science Songs
   http://www.learninggamesforkids.com/science_songs.html

The Magic School Bus™
   http://www.scholastic.com/magicschoolbus/

NASA (for students)
   http://www.nasa.gov/audience/forstudents/

NASA/MSU-Bozeman CERES Project "MoonQuest"
   http://btc.montana.edu/ceres/html/Quem.html

National Geographic for Kids
   http://kids.nationalgeographic.com/kids/

National Wildlife Federation© for Kids
   http://www.nwf.org/Kids.aspx

Newton's Apple
   http://www.tpt.org/newtons/

PBS: NOVA®
   http://www.pbs.org/wgbh/nova/programs/

Planet Science
   http://www.planet-science.com/

School House Rock
   http://www.schoolhouserock.tv/

Science Is Fun
   http://www.scifun.org/

Science Kids©
   http://www.sciencekids.co.nz/

Science News for Kids
   http://www.sciencenewsforkids.org/

Scientific American™
   http://www.scientificamerican.com/

Siemens Foundation Competition
   http://www.siemens-foundation.org/en/competition.htm

Smithsonian National Museum of Natural History
   http://forces.si.edu/educators.html

Teacher's Domain
   http://www.teachersdomain.org/

TeacherTube®
   http://www.teachertube.com/

TIME® for Kids
   http://www.timeforkids.com/

Toshiba ExploraVision
   http://www.exploravision.org/

United States Geological Survey (USGS) for Kids
   http://earthquake.usgs.gov/learn/kids/

YouTube™
   http://www.youtube.com

# Project-Based Assessment Rubric

**Project:** Write a song comparing the inner and outer planets.

**Lesson Objective:** Students compare the inner and outer planets.

|  | 4 | 3 | 2 | 1 |
|---|---|---|---|---|
| **Content** | The lyrics consistently and thoroughly compared the inner and outer planets. | The lyrics satisfactorily compared the inner and outer planets. | The lyrics compared the inner and outer planets with some unrelated content. | The lyrics unsatisfactorily compared the inner and outer planets. |
| **Format** | The lyrics fully matched a recognizable melody. | The lyrics mostly matched a recognizable melody. | The lyrics somewhat matched a recognizable melody. | The melody was unrecognizable. |
| **Format** | Ideas flowed easily from one to the next. | Ideas moved somewhat easily from one to the next. | Ideas moved haltingly from one to the next. | Ideas were disorganized. |
| **Elaboration** | The song included at least three correct comparisons between the inner and outer planets. | The song included two correct comparisons between the inner and outer planets. | The song included one correct comparison between the inner and outer planets. | The song included no correct comparisons between the inner and outer planets. |
| **Elaboration** | The song used specific vocabulary. | The song used some specific vocabulary. | The song used little specific vocabulary. | The song used no specific vocabulary. |
| **Conventions** | The song followed correct grammar usage. | The song followed mostly correct grammar. | The song followed some correct grammar. | The song had frequent grammatical errors. |
| **Conventions** | All the words were spelled correctly. | Most words were spelled correctly. | There were frequent misspellings, but it was still readable. | There were frequent misspellings that made it unreadable. |
| **Overall Assessment** | The project was neat and represented the student's best work. | The project was mostly neat and mostly represented the student's best work. | The project lacked neatness and the work was only somewhat satisfactory. | The project was messy and did not represent satisfactory work. |

# Student Self-Assessment Rubric

**Name:** _____     **Date:** _____

**Unit:** _____

How did you do during this unit in science?

**1.** I completed my labs on time.        Yes        No

**2.** I completed my classwork on time.     Yes        No

**3.** I read my assignments.           Yes        No

**4.** I participated in group projects.      Yes        No

**5.** I participated in class discussions.    Yes        No

**6.** I always did my best work.         Yes        No

**7.** My projects and quizzes/tests show that I learned _____ of the information during this unit.

❏ All    ❏ Some    ❏ Most    ❏ Little/None

**8.** I believe I earned a grade of:     **A**     **B**     **C**     **D**     **F**

Here's why:

_____

_____

**9.** Something new I learned about this topic was:

_____

_____

# Two-Column Chart

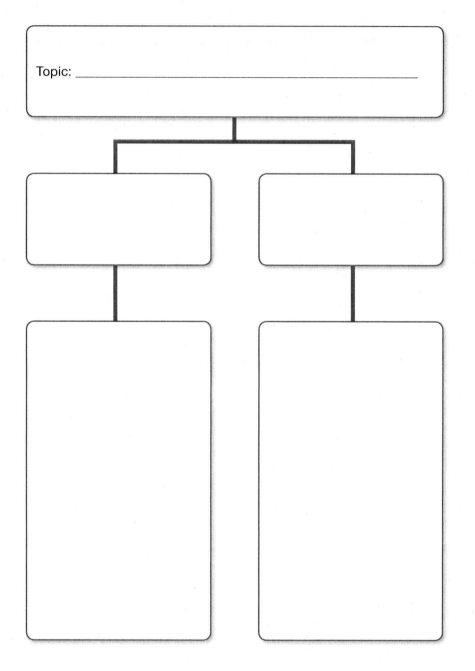

Topic: _____

# Vocabulary Game Pyramid Outline

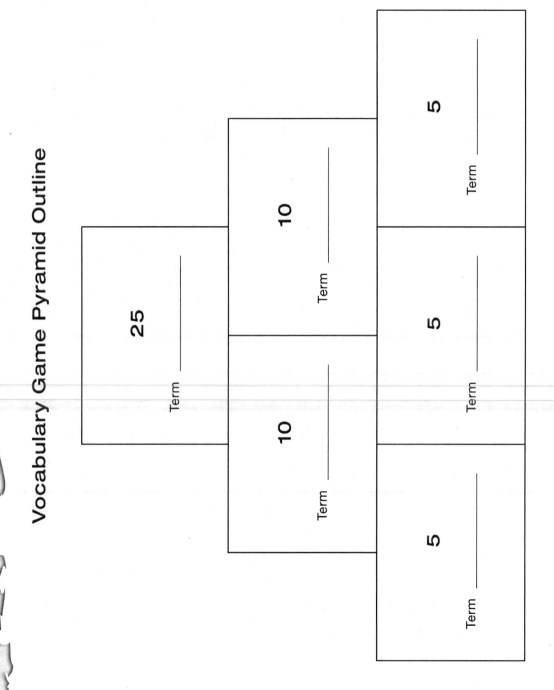

25
Term _____

10
Term _____

10
Term _____

5
Term _____

5
Term _____

5
Term _____

# A Checklist to Determine Student Mastery of the Nature of Science Objectives

As students perform the task, the teacher checks off the observed skills. For students who do not master all skills, the teacher can reassess the student using the same checklist at a later time with a different investigation.

## Observation Checklist: The Nature of Science

Name: _____

❏ Defines a problem.

❏ Plans and executes a scientific investigation.

❏ Identifies variables.

❏ Identifies the control.

❏ Collects and organizes data.

❏ Interprets data in charts, tables, and graphs.

❏ Analyzes information.

❏ Makes predictions.

❏ Defends conclusions.

# Choices Board for a Unit on Animal Studies

**Directions:** Choose any three tasks to make "Tic-Tac-Toe."

| | | |
|---|---|---|
| Fold a paper in half. Draw and describe on the left how one animal looks in the summer and how this same animal looks in the winter on the right. | List at least five ways animals camouflage themselves from predators or prey. | Write to explain: If you were an animal hiding from a predator, how would you do so? Describe how this helps you. |
| Write to explain: If you were an animal hiding from prey, how would you do so? Describe how this helps you. | Draw one environment: forest, desert, arctic, or ocean. Include at least five animals that are camouflaged in this environment. | Make a collage using torn, colored paper. Your collage should be of a camouflaged predator and its prey in the same environment. |
| List at least five ways an animal could camouflage itself in your house. | What is the best animal defense against predators? Explain why you think this. | Write a story from the perspective of a prey species: "My Great Escape from My Number One Enemy." |

# A 5E Lesson Plan Outline

| Topic: | | Grade Level: | Class: |
|---|---|---|---|
| Objective(s): | | | |

| Engage | Activity: | Materials: |
|---|---|---|
| | Formative Assessment Opportunity | |

| Explore | Activity: | Materials: |
|---|---|---|
| | Formative Assessment Opportunity | |

| Explain | Activity: | Materials: |
|---|---|---|
| | Formative Assessment Opportunity | |

| Elaborate or Extend | Activity: | Materials: |
|---|---|---|
| | Formative Assessment Opportunity | |

| Evaluate | Activity: | Materials: |
|---|---|---|
| | Summative Assessment Opportunity | |

# References Cited

Azzam, Amy. 2008. "Clash! The World of Debate." *Educational Leadership* 65 (5): 68–72.

Beck, Isabel, Margaret McKeown, and Linda Kucan. 2002. *Bringing Words to Life: Robust Vocabulary Instruction*. New York: The Guilford Press.

Bell, Randy, Lara Smetana, and Ian Binns. 2005. "Simplifying Inquiry Instruction." *The Science Teacher* (October): 30–33.

Bransford, John, Ann L. Brown, and Rodney R. Cocking, eds. 2000. *How People Learn: Brain, Mind, Experience, and School*. Washington, DC: National Academy Press.

Bybee, Rodger W., Joseph A. Taylor, April Gardner, Pamela Van Scotter, Janet Carlson Powell, Anne Westbrook, and Nancy Landes. 2006. *BSCS 5E Instructional Model: Origins and Effectiveness*. Colorado Springs, CO: BSCS.

Common Core State Standards. 2012. Accessed October 15, 2012. http://www.corestandards.org/.

Crowther, Gregory. 2012. "Using Science Songs to Enhance Learning: An Interdisciplinary Approach." *CBE Life Sciences Education*. Spring. 11 (1): 26–30.

Echevarría, Jana, MaryEllen Vogt, and Deborah J. Short. 2008. *Making Content Comprehensible for English Learners: The SIOP Model*. Boston, MA: Pearson.

Fountas, Irene C., and Gay Pinnell. 2001. *Guiding Readers and Writers, Grades 3–6: Teaching Comprehension, Genre, and Content Literacy*. Portsmouth, NH: Heinemann.

Harvey, Stephanie. 1998. *Nonfiction Matters: Reading, Writing, and Research in Grades 3–8*. Portland, ME: Stenhouse Publishers.

Herron, Marshall. 1971. "The Nature of Scientific Enquiry." *School Review* 79 (2): 171–212.

Hunter, Madeline. 1982. *Mastery Teaching: Increasing Instructional Effectiveness in Elementary and Secondary Schools*. Thousand Oaks, CA: Corwin Press, Inc.

Jensen, Eric. 2008. *Brain-Based Learning: The New Paradigm of Teaching*. Thousand Oaks, CA: Corwin Press.

Lemke, Cheryl, and Ed Coughlin. 2009. "The Change Agents." *Educational Leadership* 67 (1): 54–59.

Marzano, Robert J. 2004. *Building Background Knowledge for Academic Achievement: Research on What Works in Schools*. Alexandria, VA: Association for Supervision and Curriculum Development.

———. 2007. *The Art and Science of Teaching: A Comprehensive Framework for Effective Instruction*. Alexandria, VA: Association for Supervision and Curriculum Development.

Marzano, Robert J., and Debra J. Pickering. 2011. *The Highly Engaged Classroom*. Bloomington, IN: Marzano Research Laboratory.

Marzano, Robert J., Debra J. Pickering, and Jane E. Pollock. 2001. *Classroom Instruction that Works: Research-based Strategies for Increasing Student Achievement*. Alexandria, VA: Association for Supervision and Curriculum Development.

McGregor, Tanny. 2007. *Comprehension Connections: Bridges to Strategic Reading*. Portsmouth, NH: Heinemann.

Morin, David. 2008. "Why Did the Chicken Cross the Road?" Harvard University Department of Physics. http://www.physics.harvard.edu/academics/undergrad/chickenroad.html.

National Science Board. 2007. "National Action Plan for Addressing the Critical Needs of the U.S. Science, Technology, Engineering, and Mathematics Education System." Arlington, VA: National Science Foundation.

National Science Education Standards. 1996. Washington, DC: The National Academies Press. http://www.nap.edu/catalog/4962.html.

National Science Teachers Association. 2002. *An NSTA Position Statement: Elementary School Science*. Arlington, VA: NSTA.

Next Generation Science Standards. 2012. Accessed October 15, 2012. http://www.nextgenscience.org/.

Ogle, Donna M. 1986. "K-W-L: A Teaching Model that Develops Active Reading of Expository Text." *Reading Teacher* 39: 564–570.

Organisation for Economic Co-operation and Development. 2004. "Problem Solving for Tomorrow's World—First Measures of Cross-Curricular Competencies from PISA 2003." Accessed October 15, 2012. http://www.oecd.org/pisa/pisaproducts /pisa2003/.

Partnership for 21st Century Skills. 2011. Accessed October 15, 2012. http://www.p21.org/.

Rasinski, Timothy, Nancy Padak, Rick Newton, and Evangeline Newton. 2008. *Greek & Latin Roots: Keys to Building Vocabulary*. Huntington Beach, CA: Shell Education.

Sagan, Carl. 1986. *Boca's Brain: Reflections on the Romance of Science*. New York: Random House.

Schlechty Center. 2009. Accessed October 15, 2012. http://www.schlechtycenter.org/.

Schurr, Sandra. 2012. *Authentic Assessment: Active, Engaging Product and Performance Measures*. Westerville, OH: Association for Middle Level Education.

Tomlinson, Carol Ann. 1999. *The Differentiated Classroom: Responding to the Needs of All Learners*. Alexandria, VA: Association for Supervision and Curriculum Development.

Trefil, James, and Wanda O'Brien-Trefil. 2009. "The Science Students Need to Know." *Educational Leadership* 67 (1): 28–33.

Wiggins, Grant, and Jay McTighe. 1998. *Understanding by Design*. Alexandria, VA: ASCD.